Communications
in Computer and Information Science   2556

Series Editors

Gang Li , *School of Information Technology, Deakin University, Burwood, VIC, Australia*
Joaquim Filipe , *Polytechnic Institute of Setúbal, Setúbal, Portugal*
Zhiwei Xu, *Chinese Academy of Sciences, Beijing, China*

Rationale

The CCIS series is devoted to the publication of proceedings of computer science conferences. Its aim is to efficiently disseminate original research results in informatics in printed and electronic form. While the focus is on publication of peer-reviewed full papers presenting mature work, inclusion of reviewed short papers reporting on work in progress is welcome, too. Besides globally relevant meetings with internationally representative program committees guaranteeing a strict peer-reviewing and paper selection process, conferences run by societies or of high regional or national relevance are also considered for publication.

Topics

The topical scope of CCIS spans the entire spectrum of informatics ranging from foundational topics in the theory of computing to information and communications science and technology and a broad variety of interdisciplinary application fields.

Information for Volume Editors and Authors

Publication in CCIS is free of charge. No royalties are paid, however, we offer registered conference participants temporary free access to the online version of the conference proceedings on SpringerLink (http://link.springer.com) by means of an http referrer from the conference website and/or a number of complimentary printed copies, as specified in the official acceptance email of the event.

CCIS proceedings can be published in time for distribution at conferences or as post-proceedings, and delivered in the form of printed books and/or electronically as USBs and/or e-content licenses for accessing proceedings at SpringerLink. Furthermore, CCIS proceedings are included in the CCIS electronic book series hosted in the SpringerLink digital library at http://link.springer.com/bookseries/7899. Conferences publishing in CCIS are allowed to use Online Conference Service (OCS) for managing the whole proceedings lifecycle (from submission and reviewing to preparing for publication) free of charge.

Publication process

The language of publication is exclusively English. Authors publishing in CCIS have to sign the Springer CCIS copyright transfer form, however, they are free to use their material published in CCIS for substantially changed, more elaborate subsequent publications elsewhere. For the preparation of the camera-ready papers/files, authors have to strictly adhere to the Springer CCIS Authors' Instructions and are strongly encouraged to use the CCIS LaTeX style files or templates.

Abstracting/Indexing

CCIS is abstracted/indexed in DBLP, Google Scholar, EI-Compendex, Mathematical Reviews, SCImago, Scopus. CCIS volumes are also submitted for the inclusion in ISI Proceedings.

How to start

To start the evaluation of your proposal for inclusion in the CCIS series, please send an e-mail to ccis@springer.com.

Henry Han · James Stamey
Editors

New Frontiers in Data Science

4th Southwest Data Science Conference, SDSC 2025
Waco, TX, USA, March 21–22, 2025
Revised Selected Papers

Editors
Henry Han
Baylor University
Waco, TX, USA

James Stamey
Baylor University
Waco, TX, USA

ISSN 1865-0929   ISSN 1865-0937 (electronic)
Communications in Computer and Information Science
ISBN 978-3-031-99878-2   ISBN 978-3-031-99879-9 (eBook)
https://doi.org/10.1007/978-3-031-99879-9

© The Editor(s) (if applicable) and The Author(s), under exclusive license to Springer Nature Switzerland AG 2026

This work is subject to copyright. All rights are solely and exclusively licensed by the Publisher, whether the whole or part of the material is concerned, specifically the rights of translation, reprinting, reuse of illustrations, recitation, broadcasting, reproduction on microfilms or in any other physical way, and transmission or information storage and retrieval, electronic adaptation, computer software, or by similar or dissimilar methodology now known or hereafter developed.
The use of general descriptive names, registered names, trademarks, service marks, etc. in this publication does not imply, even in the absence of a specific statement, that such names are exempt from the relevant protective laws and regulations and therefore free for general use.
The publisher, the authors and the editors are safe to assume that the advice and information in this book are believed to be true and accurate at the date of publication. Neither the publisher nor the authors or the editors give a warranty, expressed or implied, with respect to the material contained herein or for any errors or omissions that may have been made. The publisher remains neutral with regard to jurisdictional claims in published maps and institutional affiliations.

This Springer imprint is published by the registered company Springer Nature Switzerland AG
The registered company address is: Gewerbestrasse 11, 6330 Cham, Switzerland

If disposing of this product, please recycle the paper.

Preface

Data science has evolved from an exploratory craft into a transformative engine—one that now shapes public policy, drives medical breakthroughs, powers algorithmic trading, optimizes global supply chains, and guides the myriad algorithmic judgments woven into our daily routines. Fueling this evolution is a new generation of AI components: foundation models boasting near-human fluency, self-supervised representation learners that extract latent structure from unlabeled data, and autonomous agents capable of coordinating complex, real-time workflows. These innovations have become the nervous system of modern analytics pipelines, turning raw data into actionable insight at unprecedented scale and speed. Yet with such power comes an imperative as pressing as any technical advance: embedding ethics and accountability throughout the entire data-to-decision lifecycle—from data collection and feature engineering to model deployment, monitoring, and eventual decommissioning—so that progress benefits society equitably rather than magnifying existing disparities.

Over the past decade, the community has celebrated remarkable AI advances: large-language models (LLMs) that achieve near-human fluency; diffusion architectures that rival professional artists; vision systems that match or exceed radiologists' performance; and deep-reinforcement learners that navigate both markets and robotic tasks. Each triumph, however, has revealed parallel vulnerabilities: coded social biases that exacerbate inequity; "black-box" architectures that elude meaningful audit; machine-learning assessments that falter on certain real-world datasets; reproducibility gaps tied to non-deterministic GPU kernels; privacy threats arising from trillion-parameter memorization; and catastrophic failure modes when autonomous agents drift beyond their training distributions.

The era of ethical, AI-driven data science is coming because it must—without it, progress risks entrenching the very disparities it seeks to resolve. For example, traditional attention mechanisms can introduce bias in speech-emotion and image-recognition tasks; without careful domain-specific tuning, these modules may reinforce unfair outcomes. In journalism, legacy clustering and vectorization methods—such as raw TF-IDF on newsroom comments—can mute minority voices and distort the embedding space. Optimization models, too, can go awry: a poorly designed deep-reinforcement learner might sacrifice both ethics and profitability in operational decisions. Moreover, the underlying data-science infrastructure itself must evolve: current Python-centric AI tooling, while powerful, struggles to accommodate ever-increasing data volumes and compute demands. Emerging languages and frameworks—Mojo among them—promise to reshape how we build high-performance, ethically guided pipelines.

Addressing these challenges requires principled evaluation metrics, transparent model architectures, and reproducible workflows anchored in fairness, accountability, and societal impact. It means rethinking every stage—data ingestion, model design, deployment, and monitoring—through an ethical lens, from initial feature engineering to final decision audits.

Infrastructure, too, is an ethical choice. Numerical nondeterminism in GPU kernels can shatter reproducibility guarantees; opaque, cloud-only runtimes can obscure audit trails; and energy-hungry training regimes shift environmental costs onto already disadvantaged communities. Selecting or designing toolchains that provide transparent numerics, deterministic execution, and resource-aware optimization is therefore not merely an engineering preference but an ethical obligation. Emerging frameworks such as Mojo, whose open compilation flow and low-level control enable both speed and exact repeatability, show that performance and principled stewardship can coexist—and must, if data-science pipelines are to remain trustworthy from first byte to final decision.

Each contribution in this volume shows how cutting-edge AI methods—ethical newsroom clustering, grey-system forecasting, dueling-network reinforcement learning, explainable KAN-based medical-image analysis, Mojo-accelerated k-nearest-neighbor learning with SIMD, adaptive-attention models for speech-emotion recognition, and more—can be fused with rigorous operational frameworks to deliver decisions that are not only efficient and profitable but also transparent, auditable, and just. We hope these pages inspire practitioners and scholars alike to push the frontier where algorithmic ingenuity and ethical stewardship advance in concert, ensuring that the next wave of data-science innovation uplifts every stakeholder it touches.

This book constitutes the proceedings of the Fourth Southwest Data Science Conference (SDSC 2025), which took place in Waco, Texas, USA, on March 21–22, 2025. All papers in this volume were selected from 64 submissions that underwent a rigorous single-blind review process, with each paper evaluated by at least three expert reviewers. Of those submissions, 11 high-quality full papers were accepted for inclusion.

May 2025

Henry Han
James Stamey

Organization

General Chair

Henry Han   Baylor University, USA

Program Committee Chairs

Henry Han   Baylor University, USA
James Stamey   Baylor University, USA

Steering Committee

Henry Han   Baylor University, USA
James Stamey   Baylor University, USA
Erich Baker   Belmont University, USA
Greg Hamerly   Baylor University, USA
Tim Sheng   Baylor University, USA
Hyeong-Moo Shin   Baylor University, USA
Amanda Hering   Baylor University, USA
Qiang Wu   University of Tennessee, Knoxville, USA

Program Committee

Hisham Al-Mubaid   University of Houston, USA
Andrew Chen   University of Kansas, USA
Haihua Chen   University of North Texas, USA
Erdogan Dogdu   Angelo State University, USA
Jeff Forest   Slippery Rock University, USA
Mark Ferguson   University of South Carolina, USA
Michael Gallaugher   Baylor University, USA
Chan Gu   Ball State University, USA
Keith Hubbard   Stephen F. Austin State University
Xiuzhen Hu   Inner Mongolia University of Technology, China
David Kahle   Baylor University, USA
Haiquan Li   University of Arizona, USA

Dondong Li	South University of Technology, China
David Li	Institute of Plant Physiology and Ecology, CAS, China
Huiming Liu	Guangxi University, USA
Wenbin Liu	Guangzhou University, China
Michael Pokojovy	Old Dominion University, USA
Eli Olinick	Southern Methodist University, USA
Guimin Qin	Xidian University, China
Jianhua Ruan	University of Texas at San Antonio, USA
Greg Speegle	Baylor University, USA
Yang Sun	Park University, USA
Ye Tian	Case Western Reserve University, USA
Stellar Tao	UT Health Center, USA
Tie Wei	Guangxi University, China
Jiacun Wang	Monmouth University, USA
Yi Wu	New York University, USA
Juanying Xie	Shaanxi Normal University, China
Honggang Zhang	University of Massachusetts, Boston, USA
Liang Zhao	University of Michigan, USA
Jeff Zhang	California State University, Northridge, USA
Joe Zhou	Columbia University, USA

Contents

Ethical AI-Driven Data Science

Explainable Diagnosis of Alzheimer's Disease Using Graph Kolmogorov-Arnold Networks 3
    Tianqi Ding, Dawei Xiang, Keith E. Schubert, and Liang Dong

Ethical Bytes in Newsroom: Mapping AI's Future in Journalism 16
    Ashley Han and Henry Han

New Data-Science Infrastructure and Platforms

Unleashing Mojo: Accelerating K-Nearest Neighbor Learning 39
    Sumanth Kolli, Chujiang Wu, and Henry Han

A Practical Comparison of Bayesian Computing Platforms in R 58
    Evan Miyakawa and David Kahle

Attention-Enhanced Deep Learning

CATC-Net: A CoAttention-Guided Temporal Capsule Network for Speech Emotion Recognition 77
    Yuanyuan Wei and Heming Huang

YOLOv5-TS: Channel-Spatial Attention Guided Bidirectional Feature Fusion for Traffic Sign Detection 92
    Yanbang Zhang, Fen Zhang, Yuanfan Gao, and Enyan Guo

Biomedical and Educational Data Science

Epidemiological and Transcriptomic Analysis of a Multidrug-Resistant *Pseudomonas Aeruginosa* Strain 107
    Yao Zhou, Runqing Shi, Mengshan Zhou, Yuting Zhang, Shuo Wang, Yan Song, and Yaodong Chen

Feature Engineering on LMS Data to Optimize Student Performance Prediction 125
    Keith Hubbard and Sheilla Amponsah

Socioeconomic Effects on Health and Well-Being Using U.S. County-Level Data 143
Lily Shaw, Erdogan Dogdu, Roya Choupani, Steven Womack, and Minh Le

AI-Driven Supply Chain and Operations Management

Automatic Pricing and Replenishment Decisions for Vegetable Products Based on Grey Prediction Model and 0-1 Programming 163
Yanyan Xue, Yuanyuan Zheng, Jianwei Xiao, and Xuemei Yang

Optimization of Multi-factory Remanufacturing Process with Drone Delivery Using Dueling DQN 184
Yingjun Ji, Shaokang Dai, Xiwang Guo, Jiacun Wang, Shujin Qin, and Bella Wu

Author Index 201

Ethical AI-Driven Data Science

Explainable Diagnosis of Alzheimer's Disease Using Graph Kolmogorov-Arnold Networks

Tianqi Ding[1], Dawei Xiang[2], Keith E. Schubert[1], and Liang Dong[1](✉)

[1] Baylor University, Waco, TX 76798, USA
liang_dong@baylor.edu
[2] University of Connecticut, Storrs, CT 06066, USA

Abstract. Alzheimer's Disease (AD) is a progressive neurodegenerative disorder that poses significant diagnostic challenges due to its complex etiology. Graph Convolutional Networks (GCNs) have shown promise in modeling brain connectivity for AD diagnosis, yet their reliance on linear transformations limits their ability to capture intricate nonlinear patterns in neuroimaging data. To address this, we propose GCN-KAN, an architecture that integrates Kolmogorov-Arnold Networks (KANs) into GCNs to enhance both diagnostic accuracy and interpretability. Leveraging structural MRI data from 91 subjects, our model employs learnable spline-based transformations to better represent brain region interactions. Evaluated on the Alzheimer's Disease Neuroimaging Initiative (ADNI) dataset, GCN-KAN outperforms traditional GCNs by 5.2% in classification accuracy (62.6% vs. 57.4%) while providing interpretable insights into key brain regions associated with AD. This approach offers a robust and explainable tool for AD diagnosis, potentially facilitating earlier intervention and more personalized treatment planning.

Keywords: Graph Convolutional Network · Kolmogorov-Arnold Network · Alzheimer's Disease · Machine Learning · Explainable AI · Neuroimaging

1 Introduction

Alzheimer's Disease (AD) represents a significant public health challenge as the leading cause of dementia worldwide, characterized by progressive memory loss, cognitive decline, and neurodegeneration [1]. Despite advances in neuroimaging techniques, early and accurate diagnosis remains challenging due to AD's complex etiology. Structural magnetic resonance imaging (MRI) reveals patterns of brain atrophy, particularly in the hippocampus and entorhinal cortex, which correlate with cognitive decline and often precede clinical symptoms by years. However, translating these neuroimaging biomarkers into accurate diagnostic tools requires computational approaches that can effectively model the complex relationships between brain structures.

Graph-based deep learning methods have emerged as powerful tools for modeling brain networks by representing regions of interest (ROIs) as nodes and their interactions as edges [14]. Graph Convolutional Networks (GCNs) have garnered particular attention for their ability to automatically learn features from brain network topology [12]. These approaches can capture spatial dependencies and structural patterns critical for differentiating between healthy and pathological brain states. However, conventional GCN architectures rely on fixed linear transformations followed by simple nonlinear activations, which may inadequately capture the intricate nonlinear relationships underlying AD pathology. While multi-modal GCN approaches have been explored for integrating MRI with PET and genetic data [21], they typically require additional mechanisms for feature fusion and interpretability [16,22], while still inheriting the fundamental linear limitations of standard GCN layers.

The recently introduced Kolmogorov-Arnold Networks (KANs) offer a promising alternative by replacing linear weight matrices with learnable spline-based functions [8]. This architecture, inspired by the Kolmogorov-Arnold representation theorem, demonstrates enhanced capacity for modeling complex nonlinear patterns while maintaining interpretability. By applying learnable univariate functions directly to edge features, KANs can represent a wider range of functional relationships than traditional neural networks with fixed activation functions. Integrating KAN principles into a GCN framework could potentially overcome the expressiveness limitations of traditional GCNs while preserving their ability to model spatial dependencies in brain networks [6,17].

In this paper, we introduce GCN-KAN, a novel hybrid model that combines the spatial learning capabilities of GCNs with the enhanced nonlinear representation power of KANs. Our approach focuses on single-modal analysis of structural MRI data, prioritizing interpretability and clinical feasibility. The GCN-KAN framework enhances AD diagnosis accuracy while providing explainable insights into the neuroanatomical bases of classification decisions. Our contributions are threefold: (1) we propose a novel graph-based approach for AD diagnosis leveraging MRI-derived brain connectivity that incorporates learnable spline-based transformations to enhance GCN expressiveness; (2) our model achieves 5.2% improvement in classification accuracy (62.6% vs. 57.4%) over traditional GCN baselines on the Alzheimer's Disease Neuroimaging Initiative (ADNI) dataset; and (3) we enhance interpretability by identifying critical brain regions such as the hippocampus, inferior parietal gyrus, and amygdala that align with established clinical findings in the neurological literature. This integration of advanced graph-based learning with interpretable nonlinear transformations offers a promising approach for explainable AD diagnosis that balances performance with clinical trustworthiness.

2 Related Work

2.1 Alzheimer's Disease Neuroimaging

Alzheimer's Disease manifests through progressive neurodegeneration that begins years before clinical symptoms appear [1]. Structural MRI can detect early patterns of brain atrophy, particularly in medial temporal structures like the hippocampus and entorhinal cortex, offering biomarkers for early detection. Modern approaches model the brain as a network where regions interact through structural or functional connections, aligning with the understanding of AD as a disconnection syndrome affecting entire neural circuits rather than isolated regions. This network perspective enables more comprehensive analysis of the complex pathophysiological changes underlying AD progression.

2.2 Graph Neural Networks for Brain Analysis

Graph Convolutional Networks (GCNs) have become instrumental in modeling brain connectivity for neurological disorders [11]. For a graph with adjacency matrix $A \in \mathbb{R}^{N \times N}$ and feature matrix $X \in \mathbb{R}^{N \times F}$, the GCN layer is formulated as

$$H^{(l+1)} = \sigma\left(\tilde{D}^{-1/2}\tilde{A}\tilde{D}^{-1/2}H^{(l)}W^{(l)}\right) \quad (1)$$

where $H^{(l)}$ is the node feature matrix at layer l, $\tilde{A} = A + I$ is the adjacency matrix with self-loops, $\tilde{D}$ is the degree matrix of $\tilde{A}$, $W^{(l)} \in \mathbb{R}^{F^{(l)} \times F^{(l+1)}}$ is a learnable weight matrix, and σ is a nonlinear activation function [7].

In AD research, Zhou et al. [21] developed a sparse interpretable GCN for multi-modal integration of neuroimaging data. Their approach introduced importance probabilities for features and edges, enhancing model interpretability. However, despite these advances, multi-modal approaches increase complexity while the fundamental linear limitations of GCNs persist. Several studies have attempted to enhance GCN expressiveness through attention mechanisms [4], but these often reduce interpretability, which is crucial in clinical applications.

2.3 Kolmogorov-Arnold Networks

Kolmogorov-Arnold Networks (KANs) represent a significant innovation in neural network architecture, drawing inspiration from the Kolmogorov-Arnold representation theorem [6,17]. This theorem states that any continuous multivariate function $f : [0,1]^n \to \mathbb{R}$ can be expressed as

$$f(x_1, x_2, \ldots, x_n) = \sum_{q=1}^{2n+1} \Phi_q \left(\sum_{p=1}^{n} \phi_{q,p}(x_p)\right) \quad (2)$$

where Φ_q and $\phi_{q,p}$ are continuous univariate functions.

Unlike traditional multilayer perceptrons that apply linear transformations followed by fixed nonlinear activations, KANs employ learnable spline-based functions directly on individual features. For input features $x = [x_1, x_2, \ldots, x_n]$, a single layer KAN transformation computes

$$z_j = \sum_{i=1}^{n} \phi_{j,i}(x_i) \quad (3)$$

where each $\phi_{j,i}$ is a distinct univariate function applied to feature x_i.

These functions are typically parameterized as piecewise linear splines with learnable coefficients

$$\phi(x) = w_b b(x) + w_s \sum_{k=1}^{K} c_k B_k(x) \tag{4}$$

where $b(x)$ is a basis function (typically SiLU), $B_k(x)$ are B-spline basis functions defined over a grid of knot points, w_b and w_s are learned weights, and c_k are learnable coefficients.

Recent work has explored integrating KAN principles into graph-based models [6,17], demonstrating improved performance on node classification tasks. However, applications of KANs to neuroimaging and specifically AD diagnosis remain largely unexplored. Our work bridges this gap by combining the spatial learning capabilities of GCNs with the enhanced nonlinear representation power of KANs for improved AD diagnosis.

3 Methodology

3.1 Graph Construction

We construct brain connectivity graphs using structural MRI data from the ADNI dataset [9], processed to extract 90 ROIs based on the Automated Anatomical Labeling (AAL) atlas [13]. Each node corresponds to an ROI, with feature vector $x_i \in \mathbb{R}$ representing the normalized gray matter volume derived from voxel-based morphometry (VBM)—a well-established biomarker for AD [2]. The feature matrix $X \in \mathbb{R}^{90 \times 1}$ comprises these volumetric measurements for all ROIs.

We define the weighted adjacency matrix $A \in \mathbb{R}^{90 \times 90}$ using thresholded Pearson correlation between ROI features

$$A_{i,j} = \begin{cases} \text{corr}(x_i, x_j) & \text{if } |\text{corr}(x_i, x_j)| > \tau \\ 0 & \text{otherwise} \end{cases} \tag{5}$$

where $\text{corr}(x_i, x_j)$ computes the Pearson correlation coefficient between features of regions i and j, and $\tau = 0.1$ serves as a sparsification threshold. This thresholding operation preserves significant connections while reducing noise and computational complexity, yielding a sparse representation that better captures the underlying neuroanatomical relationships [15].

3.2 GCN-KAN Architecture

The GCN-KAN model integrates Graph Convolutional Network (GCN) layers with Kolmogorov-Arnold Network (KAN) layers to enhance expressivity while maintaining interpretability. Figure 1 illustrates the overall architecture, which

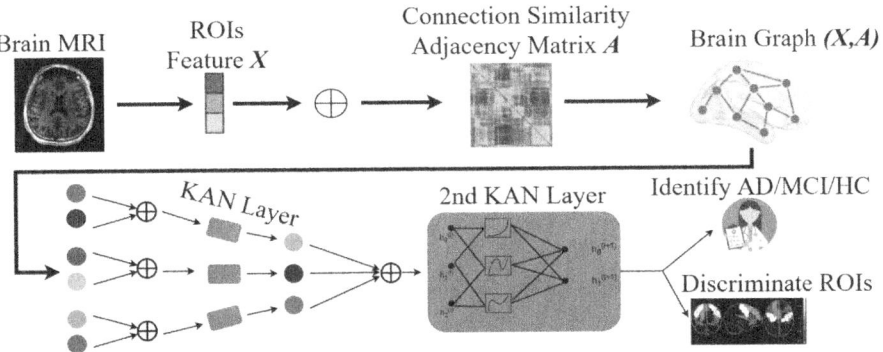

Fig. 1. Overall architecture of our proposed GCN-KAN model for Alzheimer's disease diagnosis. The framework integrates GCN layers for capturing spatial dependencies across brain regions with KAN layers that implement learnable spline-based transformations. This hybrid approach enhances nonlinear representation capabilities while maintaining interpretability, enabling identification of neuroanatomically relevant biomarkers. The model processes structural MRI features through sequential GCN and KAN layers before applying global pooling and classification to generate diagnostic predictions.

consists of four main components: GCN layers for spatial learning, KAN layers for nonlinear transformations, a pooling operation, and a classification layer.

Our model adopts a structure where KAN layers are applied after each GCN layer. This design choice provides several advantages: (1) it enhances layer-wise nonlinearity, ensuring that both low-level and high-level node features benefit from adaptive spline-based transformations; (2) it enables hierarchical feature refinement by progressively enhancing the representations learned by GCN layers; and (3) it improves interpretability by providing insights into feature transformations at multiple levels of abstraction.

GCN Layers. We employ two consecutive GCN layers to learn hierarchical spatial dependencies in the brain network. For a graph with feature matrix $X \in \mathbb{R}^{90 \times 1}$ and adjacency matrix $A \in \mathbb{R}^{90 \times 90}$, the normalized graph Laplacian is computed as

$$\tilde{L} = \tilde{D}^{-\frac{1}{2}} \tilde{A} \tilde{D}^{-\frac{1}{2}} \tag{6}$$

where $\tilde{A} = A + I$ adds self-loops to ensure nodes contribute to their own representations, and $\tilde{D}$ is the diagonal degree matrix with $\tilde{D}_{ii} = \sum_j \tilde{A}_{ij}$. The first GCN layer transforms the input features into a 32-dimensional hidden representation

$$H^{(1)} = \text{ReLU}\left(\tilde{L} X W^{(1)}\right) \tag{7}$$

where $W^{(1)} \in \mathbb{R}^{1 \times 32}$ is a learnable weight matrix and $\text{ReLU}(x) = \max(0, x)$ is the nonlinear activation function. The second GCN layer further refines this

representation
$$H^{(2)} = \text{ReLU}\left(\tilde{L}H^{(1)}W^{(2)}\right) \quad (8)$$
where $W^{(2)} \in \mathbb{R}^{32 \times 32}$ is another learnable weight matrix.

KAN Layers. To enhance the model's capacity to capture complex nonlinear patterns in the brain connectivity data, we incorporate two KAN layers after the GCN layers. For each KAN layer, we first normalize the input feature matrix to the range $[0, 1]$ to ensure numerical stability

$$\hat{H}^{(l)} = \frac{H^{(l)} - \min(H^{(l)})}{\max(H^{(l)}) - \min(H^{(l)}) + \epsilon} \quad (9)$$

where $\epsilon = 10^{-8}$ is a small constant added to prevent division by zero, and $\min(H^{(l)})$ and $\max(H^{(l)})$ return the minimum and maximum values across all elements of $H^{(l)}$.

The KAN transformation implements piecewise linear splines with learnable coefficients. For each node i and layer l, the transformation is defined as

$$H_i^{(l+1)} = \sum_{j=1}^{32} \phi_{i,j}(\hat{H}_j^{(l)}) \quad (10)$$

where $\phi_{i,j}$ is a learnable univariate function applied to the j-th feature of node i. Each function $\phi_{i,j}$ is parameterized as a linear combination of basis functions

$$\phi_{i,j}(x) = \sum_{k=1}^{G} c_{i,j,k} B_k(x) \quad (11)$$

The basis functions $B_k(x)$ are defined as ReLU activations centered at fixed grid points

$$B_k(x) = \max(0, x - g_k) \quad (12)$$

where $g_k = k/G$ for $k = 0, 1, 2, \ldots, G$ are uniformly spaced grid points in $[0, 1]$, and $c_{i,j,k}$ are the learnable coefficients that define the shape of the spline functions.

We selected a grid size of $G = 10$ after careful consideration of the trade-offs between model expressivity and computational efficiency. A smaller grid (e.g., 3–5) would limit the nonlinearities that could be captured, while a larger grid (e.g., 20–50) would increase the risk of overfitting given our dataset size. The chosen value provides sufficient flexibility to model complex relationships while maintaining parameter efficiency.

This formulation allows the KAN layers to learn adaptive nonlinear transformations for each feature dimension independently. Unlike traditional neural networks with fixed activation functions, KANs can approximate arbitrary continuous functions with varying degrees of nonlinearity across different regions of the input space. This flexibility is particularly important for modeling the complex relationships in neuroimaging data, where different brain regions may exhibit distinct patterns of atrophy or functional decline in AD.

Feature Aggregation and Classification. After the second KAN layer, we apply a global max-pooling operation to aggregate node-level features into a graph-level representation

$$z = \max_{i \in \{1,2,\ldots,90\}} H_i^{(4)} \tag{13}$$

where $H^{(4)}$ denotes the output of the second KAN layer (with $H^{(1)}$ and $H^{(2)}$ coming from the first and second GCN layers, and $H^{(3)}$ from the first KAN layer), and the maximum is taken element-wise across all nodes. This operation selects the most salient features from each dimension, capturing the most discriminative patterns across the brain network.

The aggregated representation $z \in \mathbb{R}^{32}$ is then fed into a fully connected layer for binary classification

$$\hat{y} = zW^{(5)} + b \tag{14}$$

where $W^{(5)} \in \mathbb{R}^{32 \times 2}$ and $b \in \mathbb{R}^2$ are the weight matrix and bias vector, respectively. The final class probabilities are obtained by applying the softmax function

$$p(y = c|G) = \frac{\exp(\hat{y}_c)}{\sum_{c' \in \{0,1\}} \exp(\hat{y}_{c'})} \tag{15}$$

where $c \in \{0, 1\}$ represents the binary class labels (AD or non-AD).

To mitigate overfitting, we apply dropout with rate $p = 0.2$ after each KAN layer, randomly zeroing a fraction of the feature values during training. This dropout rate was specifically chosen to balance regularization without disrupting learning, which is particularly important given the large parameter count of each KAN layer (approximately $90 \times 32 \times 32 \times 10 = 921{,}600$ parameters).

3.3 Training and Optimization

We train the GCN-KAN model by minimizing the binary cross-entropy loss function

$$\mathcal{L}_{CE} = -\frac{1}{M} \sum_{m=1}^{M} [y_m \log(p_m) + (1 - y_m) \log(1 - p_m)] \tag{16}$$

where M is the batch size, $y_m \in \{0, 1\}$ is the ground truth label for sample m, and $p_m = p(y = 1|G_m)$ is the predicted probability of AD for sample m.

We apply L_2 regularization to all trainable parameters with weight decay coefficient $\lambda = 10^{-4}$, resulting in the total loss

$$\mathcal{L}_{total} = \mathcal{L}_{CE} + \lambda \sum_{\theta \in \Theta} \|\theta\|_2^2 \tag{17}$$

where Θ represents the set of all trainable parameters in the model.

The optimization is performed using the Adam optimizer with an initial learning rate of $\eta = 5 \times 10^{-4}$ and default values for the moment estimates

($\beta_1 = 0.9$, $\beta_2 = 0.999$). To improve convergence, we implement an adaptive learning rate schedule using ReduceLROnPlateau, which reduces the learning rate by a factor of 0.5 if the validation loss does not improve for 20 consecutive epochs. The minimum learning rate is set to 10^{-6}.

To prevent overfitting, we employ early stopping with a patience of 50 epochs, terminating training if the validation loss does not decrease for 50 consecutive epochs. This patience value was determined empirically to allow sufficient time for the model to converge while preventing overfitting on our limited dataset. We train the model for a maximum of 500 epochs using mini-batches of size 32. To accelerate computation and reduce memory consumption, we implement mixed-precision training using FP16 arithmetic for forward and backward passes while maintaining master weights in FP32 precision.

3.4 Interpretability Framework

A key advantage of our GCN-KAN architecture is its inherent interpretability, particularly through the analysis of the learned spline coefficients. We define an importance score for each ROI based on the magnitude of its associated coefficients in the KAN layers

$$S_i = \frac{1}{32 \cdot G} \sum_{j=1}^{32} \sum_{k=1}^{G} |c_{i,j,k}| \tag{18}$$

where $G = 10$ is the grid size, and the absolute values of coefficients are averaged across all output dimensions and grid points. This formulation assigns higher importance to ROIs that have larger coefficient magnitudes, indicating stronger influence on the model's predictions. Using absolute values allows us to capture the contribution magnitude regardless of sign.

To facilitate visualization and analysis, we normalize the importance scores to the $[0, 1]$ range

$$\hat{S}_i = \frac{S_i - \min(S)}{\max(S) - \min(S)} \tag{19}$$

These normalized scores are mapped onto the AAL atlas for spatial visualization, allowing for the identification of brain regions most critical for AD diagnosis. To create these visualizations, we use the Nilearn library to project the importance scores onto the standard AAL template, with each voxel in a given ROI assigned its corresponding importance value. The resulting spatial distribution is visualized using glass brain plots, providing an intuitive representation of the most salient regions.

This visualization approach highlights key regions such as the hippocampus, parietal cortex, and amygdala, which are known to be significantly affected in Alzheimer's Disease. Furthermore, we can visualize the learned spline functions themselves by plotting $\phi_{i,j}(x)$ for specific ROIs of interest. These visualizations offer insights into the nonlinear transformations learned by the model, revealing how different ranges of ROI features contribute to the classification decision.

The interpretability framework bridges the gap between black-box deep learning models and explainable AI systems required for clinical applications. By identifying the most informative brain regions for AD diagnosis, our approach not only provides accurate classification but also generates neuroanatomically meaningful explanations that align with clinical knowledge, potentially enhancing trust in AI-assisted diagnostic systems.

4 Experimental Results

Our experiments utilized a subset of the Alzheimer's Disease Neuroimaging Initiative (ADNI) dataset comprising 91 subjects: 45 diagnosed with Alzheimer's Disease (AD) and 46 cognitively normal (CN) controls [10]. Each structural MRI volume underwent voxel-wise normalization to the $[0,1]$ range, followed by threshold-based segmentation into 90 anatomical regions corresponding to the Automated Anatomical Labeling (AAL) atlas [13]. For each ROI, we calculated the relative volume (i.e., the count of voxels within the bin) as the node feature, effectively capturing gray matter density variations—a critical biomarker for AD. We constructed brain connectivity graphs using the thresholded Pearson correlation approach, forming sparse adjacency matrices $\mathbf{A} \in \mathbb{R}^{90 \times 90}$. Subject identifiers were extracted from MRI filenames and mapped to ADNI clinical records to validate diagnostic labels, with only subjects labeled as AD or CN retained and the MCI group excluded from this analysis.

Given the limited sample size, we employed 5-fold cross-validation [3] to ensure robust performance estimation. In this approach, the dataset was partitioned into 5 equal-sized folds, with each fold serving as the validation set once while the remaining 4 folds were used for training. For each training iteration, the model was trained with a batch size of 32 using the Adam optimizer with an initial learning rate of 5×10^{-4} and weight decay of 10^{-4}. Dropout with rate 0.2 was applied after each KAN layer to mitigate overfitting. Training efficiency was enhanced through mixed precision (PyTorch `autocast` and `GradScaler`), while convergence was optimized using a `ReduceLROnPlateau` scheduler (patience 20) and early stopping (patience 50).

We evaluated model performance using three complementary metrics: accuracy, AUC-ROC, and F1-score. Accuracy measures the proportion of correct predictions, calculated as $(TP + TN)/(TP + TN + FP + FN)$, where TP, TN, FP, and FN denote true positives, true negatives, false positives, and false negatives, respectively. The AUC-ROC evaluates the model's discriminative ability across different thresholds by plotting true positive rate ($TPR = TP/(TP + FN)$) against false positive rate ($FPR = FP/(FP + TN)$) [19,20]. The F1-score provides the harmonic mean of precision ($TP/(TP + FP)$) and recall ($TP/(TP + FN)$), offering a balanced assessment particularly valuable for datasets with class imbalance [18].

Our GCN-KAN model achieved an accuracy of 62.6% ($\pm 1.8\%$) across the 5-fold cross-validation, demonstrating a substantial 5.2% absolute improvement over the baseline GCN (57.4% $\pm 2.2\%$). Similarly, GCN-KAN outperformed the baseline in AUC-ROC (64.1% ±1.5% versus 60.3% ±2.0%) and showed a modest improvement in F1-score (0.60 ±0.02 versus 0.59 ±0.02). Table 1 summarizes these performance metrics, highlighting the consistent improvements achieved by our proposed model across all evaluation criteria.

The learning dynamics analysis revealed different convergence patterns between the two models. The GCN model exhibited faster initial convergence but higher validation loss, suggesting potential overfitting to the training data. In contrast, GCN-KAN demonstrated smoother convergence and consistently lower validation loss, indicating superior generalization capability despite its more complex architecture.

Beyond performance metrics, a key advantage of our approach lies in its interpretability. By analyzing the learned spline coefficients in the KAN layers, we identified brain regions most influential for AD classification. The hippocampus (left), inferior parietal gyrus (right), and amygdala (right) emerged with the highest normalized importance scores of 0.65, 0.61, and 0.60, respectively [5]. To provide visual interpretation of the model's decision-making process, Fig. 2 presents a brain map visualization of these salient regions, with importance scores normalized between 0 and 1. Warmer colors (yellow to red) indicate higher relevance. We specifically focused on three ROIs based on their biological relevance to Alzheimer's Disease, corresponding to AAL atlas indices 36 (Hippocampus_L), 60 (Parietal_Inf_L), and 40 (Amygdala_L). The hippocampus is essential for memory formation and spatial navigation, serving as a primary biomarker in clinical MRI analyses and one of the earliest regions to show atrophy in AD. The inferior parietal cortex supports visuospatial processing and attention, with degeneration contributing to spatial disorientation commonly observed in AD progression. The amygdala regulates emotional responses, with early atrophy linked to emotional disturbances such as anxiety and social withdrawal often seen in AD patients.

These highlighted ROIs are consistent with established neuroimaging findings, reinforcing the clinical validity of our model's interpretability. The spatial distribution supports the hypothesis that AD-related neurodegeneration is not confined to a single region but spans multiple functionally interconnected networks. Moreover, the approximate bilateral symmetry observed in several highlighted regions suggests that our model effectively captures the global pattern of neurodegeneration typical in Alzheimer's Disease, rather than overfitting to noise or isolated artifacts. The lateral views particularly emphasize involvement of the medial temporal lobe structures, which have been extensively documented

Table 1. Performance comparison between baseline GCN and our proposed GCN-KAN model on the ADNI dataset (mean ± standard deviation across 5-fold cross-validation).

Model	Accuracy (%)	AUC-ROC (%)	F1-Score
GCN (baseline)	57.4 ± 2.2	60.3 ± 2.0	0.59 ± 0.02
GCN-KAN (ours)	**62.6 ± 1.8**	**64.1 ± 1.5**	**0.60 ± 0.02**

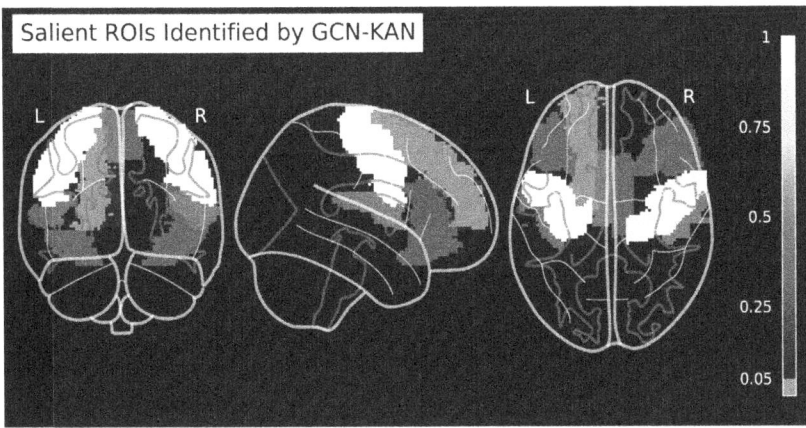

Fig. 2. Brain regions identified as most significant for Alzheimer's disease diagnosis by our GCN-KAN model. The visualization maps normalized importance scores derived from KAN layer spline coefficients onto the standard AAL atlas. Color intensity represents the relative contribution of each region to classification decisions, with warmer colors (yellow to red) indicating higher importance. Key regions with elevated importance include the hippocampus (normalized score: 0.65), inferior parietal gyrus (0.61), and amygdala (0.60)—structures known to undergo early neurodegeneration in AD pathology. The bilateral distribution with some hemispheric asymmetry aligns with established neuroimaging findings in AD progression. (Color figure online)

as hallmarks of early AD progression. This neuroanatomical relevance validates not only the model's performance but also its ability to identify biologically meaningful patterns aligned with clinical knowledge.

5 Conclusion and Future Work

This paper introduced GCN-KAN, a novel hybrid model that integrates Kolmogorov-Arnold Networks into Graph Convolutional Networks for explainable Alzheimer's Disease diagnosis from structural MRI data. By leveraging the Kolmogorov-Arnold representation theorem to incorporate learnable spline-based transformations, our approach enhances both the expressive power and interpretability of traditional GCN architectures. Experimental results on the ADNI dataset demonstrated that GCN-KAN outperforms conventional GCN models by approximately 5% in classification accuracy while providing valuable insights into neuroanatomical biomarkers of AD. The model successfully identified the hippocampus, inferior parietal gyrus, and amygdala as critical regions for diagnosis—findings that align with established clinical knowledge about AD progression.

The interpretability framework we developed enables visualization of region-specific importance scores, revealing patterns of bilateral, yet asymmetric, neurodegeneration across functionally connected brain networks. This transparent approach bridges the gap between black-box deep learning models and clinically trusted diagnostic tools, potentially facilitating adoption in medical settings where explainability is paramount. The spline-based formulation of KAN layers provides not only improved accuracy but also insights into the nonlinear relationships between brain structural changes and disease status, offering a more nuanced understanding of AD pathophysiology than linear models allow.

Despite these promising results, our work has several limitations that suggest directions for future research. The current model was developed and evaluated on a relatively small sample of 91 subjects from ADNI, which may limit its generalizability. Validation on larger, more diverse cohorts is essential to establish the robustness of our approach across different populations and scanning protocols. Furthermore, our study relies solely on structural MRI data, potentially overlooking complementary information available through other imaging modalities such as PET (for amyloid/tau deposition) or functional MRI, as well as non-imaging biomarkers from genetic and cerebrospinal fluid analyses. Extending GCN-KAN to incorporate multi-modal data could provide a more comprehensive assessment of AD and potentially boost diagnostic accuracy beyond the current performance metrics (accuracy 62.6%, AUC-ROC 64.1%, F1-score 0.60).

The modest performance, while superior to baseline methods, indicates room for architectural improvements through deeper networks, alternative spline formulations, or more sophisticated graph construction techniques that better capture the complex topology of brain connectivity in AD. Additionally, the computational overhead introduced by spline-based KAN layers may hinder scalability in large-scale or real-time clinical applications. Future work should focus on optimizing the model's efficiency through techniques such as model compression, parameter pruning, or hardware acceleration to facilitate deployment in resource-constrained clinical settings.

References

1. Breijyeh, Z., Karaman, R.: Comprehensive review on Alzheimer s disease: causes and treatment. Molecules **25**(24), 5789 (2020)
2. Busatto, G.F., Diniz, B.S., Zanetti, M.V.: Voxel-based morphometry in Alzheimer s disease. Expert Rev. Neurother. **8**(11), 1691–1702 (2008)
3. Fushiki, T.: Estimation of prediction error by using K-fold cross-validation. Stat. Comput. **21**, 137–146 (2011)
4. Hong, H., Guo, H., Lin, Y., Yang, X., Li, Z., Ye, J.: An attention-based graph neural network for heterogeneous structural learning. In: Proceedings of the AAAI Conference on Artificial Intelligence, vol. 34, pp. 4132–4139 (2020)
5. Jack, C.R., et al.: Hypothetical model of dynamic biomarkers of the Alzheimer's pathological cascade. Lancet Neurol. **9**(1), 119–128 (2010)
6. Kiamari, M., Kiamari, M., Krishnamachari, B.: GKAN: Graph kolmogorov-arnold networks. arXiv preprint arXiv:2406.06470 (2024)
7. Kipf, T.N., Welling, M.: Semi-supervised classification with graph convolutional networks. CoRR abs/ arXiv: 1609.02907 (2016)
8. Liu, Z., et al.: KAN: Kolmogorov-arnold networks. arXiv preprint arXiv:2404.19756 (2024)

9. Mueller, S.G., et al.: The Alzheimer's disease neuroimaging initiative. Neuroimaging Clinics **15**(4), 869–877 (2005)
10. Petersen, R.C., et al.: Alzheimer's disease neuroimaging initiative (ADNI) clinical characterization. Neurology **74**(3), 201–209 (2010)
11. Qi, Y., Lu, Q., Dou, S., Sun, X., Li, M., Li, Y.: Graph neural network-driven hierarchical mining for complex imbalanced data. arXiv preprint arXiv:2502.03803 (2025)
12. Rahim, N., Ahmad, N., Ullah, W., Bedi, J., Jung, Y.: Early progression detection from MCI to AD using multi-view MRI for enhanced assisted living. Image Vis. Comput. p. 105491 (2025)
13. Rolls, E.T., Huang, C.C., Lin, C.P., Feng, J., Joliot, M.: Automated anatomical labelling atlas 3. Neuroimage **206**, 116189 (2020)
14. Sang, Y., Li, W.: Classification study of Alzheimer s disease based on self-attention mechanism and DTI imaging using GCN. IEEE Access **12**, 24387–24395 (2024)
15. Tan, M., et al.: Lymonet: an advanced neck lymph node detection network for ultrasound images. IEEE J. Biomed. Health Inform. (2024)
16. Xiao, T., Zeng, L., Shi, X., Zhu, X., Wu, G.: Dual-graph learning convolutional networks for interpretable Alzheimer s disease diagnosis. In: International Conference on Medical Image Computing and Computer-Assisted Intervention. pp. 406–415. Springer (2022)
17. Zhang, F., Zhang, X.: Graphkan: enhancing feature extraction with graph Kolmogorov-Arnold networks. arXiv preprint arXiv:2406.13597 (2024)
18. Zhang, X., et al.: Multi-objective collaborative optimization algorithm for heterogeneous cooperative tasks based on conflict resolution. In: Wu, M., Niu, Y., Gu, M., Cheng, J. (eds.) ICAUS 2021. LNEE, vol. 861, pp. 2548–2557. Springer, Singapore (2022). https://doi.org/10.1007/978-981-16-9492-9_251
19. Zhang, Y., Chu, L., Xu, L., Mo, K., Kang, Z., Zhang, X.: Optimized coordination strategy for multi-aerospace systems in pick-and-place tasks by deep neural network. arXiv preprint arXiv:2412.09877 (2024)
20. Zhang, Y., et al.: Self-adaptive robust motion planning for high DoF robot manipulator using deep MPC. arXiv preprint arXiv:2407.12887 (2024)
21. Zhou, H., He, L., Chen, B.Y., Shen, L., Zhang, Y.: Multi-modal diagnosis of Alzheimer s disease using interpretable graph convolutional networks. IEEE Trans. Med. Imaging (2024)
22. Zhou, H., He, L., Zhang, Y., Shen, L., Chen, B.: Interpretable graph convolutional network of multi-modality brain imaging for Alzheimer s disease diagnosis. In: 2022 IEEE 19th International Symposium on Biomedical Imaging (ISBI), pp. 1–5. IEEE (2022)

Ethical Bytes in Newsroom: Mapping AI's Future in Journalism

Ashley Han[1] and Henry Han[2]

[1] Skyline High School, Ann Arbor, MI 48105, USA
[2] School of Engineering and Computer Science, Baylor University, Waco, TX 76798, USA
Henry_Han@baylor.edu

Abstract. While artificial intelligence (AI) offers transformative potential for journalism, its application in newsrooms presents complex ethical challenges, particularly concerning bias in analyzing unstructured text data like reader comments. This study proposes an ethical newsroom clustering framework—comprising bias-aware preprocessing, model tuning and selection, and optional consensus fusion—to mitigate such biases. It introduces a robust median silhouette score for fairness-centric and trustworthy model selection. Experiments on both synthetic data and real-world New York Times comment datasets validated the framework's efficacy, guided by the median silhouette, in preserving minority viewpoints and identifying nuanced subgroups that existing approaches obscured, thus avoiding common marginalization pitfalls. The proposed median-based silhouette score proved a more reliable and fair metric for evaluating clustering quality in noisy, real-world journalistic contexts. To the best of our knowledge, this is the first work dedicated to ethical newsroom clustering in journalism.

Keywords: AI bias · Journalism · Ethical clustering · Bias-aware preprocessing

1 Introduction

The advent of Artificial Intelligence (AI) in modern newsrooms represents a transformative milestone, akin to the arrival of the Internet itself. From automating mundane chores like comment and transcription moderation to mining vast datasets for investigative leads, different AI tools, from basic clustering, automated routines, to large language models (LLM) are rapidly becoming indispensable.

For example, newsrooms frequently leverage K-means clustering to distill thousands of reader comments into clear themes [1]. By converting comments (augmented with demographic and engagement data) into TF-IDF vectors, editors might identify segments such as long-time subscribers, intellectuals, small business owners and community activists. Reviewing each cluster's top keywords and a representative comment then enables targeted narrative angles and engagement strategies, without sifting through every remark. Similarly, the automated insights wordsmith platform transformed corporate earnings reporting at The Associated Press, freeing up roughly 20% of journalists' time and increasing output from approximately 300 manual reports to more than 3,700 automated summaries each quarter [2]. Automated report production through AI seems to be a trend in the newsroom for its efficiency and low-cost [3].

Although comprehensive metrics for LLM deployments remain limited compared to other AI tools, many news organizations now leverage models like GPT-4 to streamline workflows. A newsroom integrates an LLM-based assistant into its content management system to speed up headline generation and story outlining. The Washington Post, for example, employs a GPT-4-powered assistant to generate multiple headline variants and concise story outlines in seconds, reducing brainstorming time by nearly half [4]. Similarly, the New York Times uses AI-driven Chartbeat analytics to optimize article placement—boosting average time-on-page, and the BBC spots emerging stories with AI support [5,6]. These AI tools bring efficiency, greater media reach, and even enable innovative storytelling formats [7].

However, alongside these opportunities come serious ethical challenges in both in journalism and AI [8]. For example, when an AI system produces a flawed or misleading report or helps fake news spread faster and more broadly [9], who is responsible: developers, journalists, or publishers? Many biased results may appear in the news and reinforce existing biases because the algorithms used in handling data are themselves biased or because data collection is flawed. For example, almost all AI models or tools in newsrooms rely on historical data in which low-income and other marginalized voices are generally underrepresented, resulting in their feedback being treated as 'noise' instead of being valued. Along with other aspects, it would generate AI bias in the newsroom.

Formally, AI bias in the newsroom refers to systematic and repeatable errors within AI systems used in journalistic contexts that result in skewed perspectives, unfair representations, or inequitable outcomes, such as privileging certain groups or viewpoints while marginalizing others. Despite its critical importance, AI bias in the newsroom is not as prominently discussed within the broader discourse on AI and journalism as one might expect. How can newsrooms protect against such AI biases, which can cause or reinforce social inequities, from both an ethical and technical perspective novelly?

These challenges are essential not only for the future of journalism but also for AI and our society as a whole. AI-generated or spread fake news can blur the line between fact and fiction, erode trust in automated, AI-assisted reports, fuel divisions, derail health decision-making, and even spark social crises. Addressing AI biases in the newsroom fosters fairness and transparency in both AI and journalism, safeguards public trust in AI-based reporting, and sets a clear milestone for AI's future role in journalism.

Despite a growing body of research on AI-driven media, systematic reviews of AI's impact in journalism focus almost exclusively on high-level ethical implications and changes to newsroom workflows, offering almost no domain-specific, technical methods to detect or mitigate algorithmic bias ethically within newsroom AI systems [10]. Although the EU AI Act requires providers and deployers of generative AI systems to transparently label any AI-generated content and imposes substantial fines for non-compliance—ensuring accountability for misuse or misleading outputs from a policy-initiative perspective, it remains vague about who is responsible for fake news generated or spread with AI, not to mention measures to handle AI bias in the newsroom [11]. Beyond isolated case studies such as examinations of gender and racial stereotypes in large-language-model-generated articles—there is virtually no work that combines ethical frameworks with practical, algorithmic solutions to mitigate AI bias specifically in newsroom tools and workflows.

This paper aims to bridge this gap by focusing on AI bias within newsroom applications and proposing a concrete framework to address it. We define AI bias in the newsroom, illustrate its potential manifestations, and introduce 'ethical newsroom clustering' as a novel, trustworthy methodology to mitigate AI biases in newsroom. Our key contributions are:

1. We define and illustrate AI bias specific to newsroom contexts, particularly in the analysis of unstructured text data like reader comments, besides introducing a fairness-informed median-silhouette measure for newsroom clustering.
2. We propose 'ethical newsroom clustering,' a three-stage framework integrating (a) bias-aware preprocessing, (b) model tuning and selection, and (c) consensus fusion and quality filtering, to mitigate AI biases in the analysis of unstructured text data, thereby fostering more equitable and representative insights from reader comments and similar newsroom datasets.
3. We demonstrate the practical application and benefits of Ethical Newsroom Clustering through comparative analysis on both synthetic and real-world (New York Times comments) datasets, showcasing its ability to produce more nuanced and less biased insights.

Addressing the critical need for actionable solutions to AI bias in newsrooms, our proposed Ethical Newsroom Clustering framework offers a tangible impact and aims to provide a foundational contribution to ethical AI journalism. Its adoption can lead to more responsible deployment of AI in journalism, enhancing public trust in AI-based reporting. This, in turn, empowers journalists to leverage AI tools not just for efficiency, but as ethically-grounded aids in their mission to inform the public accurately and equitably. Furthermore, this work is anticipated to stimulate new research avenues within the burgeoning field of ethical AI journalism.

The remainder of this paper is structured as follows: Sect. 2 defines and discusses AI bias in newsrooms. Section 3 details our proposed Ethical Clustering framework, Sect. 4 introduces median silhouette score for trustworthy clustering quality evaluation. Section 5 presents our experimental setup and results. Section 6 discusses extensions and future implications. Finally, we conclude the paper.

2 Understanding AI Bias in the Newsroom

AI bias in newsrooms can be categorized into three main types: data and preprocessing bias, model and algorithm bias, and human and implementation bias (Fig. 1).

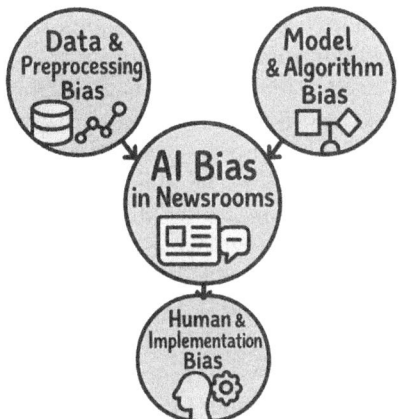

Fig. 1. Taxonomy of AI Bias in Newsrooms. This figure illustrates the three primary sources of bias: (1) data and preprocessing bias, (2) model and algorithm bias, and (3) human and implementation bias—and how they converge to shape biased outcomes in AI-driven journalistic workflows.

- *Data and preprocessing bias:* Raw data collected by newsrooms or the way it is preprocessed (e.g., during vectorization) can under-represent certain demographics or viewpoints, such as marginalized voices in comment sections, or over-represent dominant narratives. For example, a newsroom uses an AI tool to identify "expert" voices on a particular topic by scraping academic databases and established media outlets, which themselves might historically over-represent experts from certain demographics (e.g., white, male, from specific universities or organizations). AI models trained on such data inherit and may amplify these dominant voices but ignore others. Furthermore, vectorization methods may bring some biases: TF-IDF favors common terms like "economy" but downweights minority slang (e.g., "finna," "lit"), causing clustering or classification to overlook those comments and mute those voices [12].
- *Model and algorithm bias:* Different assumptions built into AI models and algorithms can introduce different biases. For example, most clustering methods often favor large, dense, and balanced groups, risking dismissal of smaller, distinct opinion groups as 'noise' or merging them into broader categories. As a result, biased reports caused by the limitations of AI algorithms and models can be viewed as generic conclusions by people.

More importantly, the assessment of AI models and algorithms can be inappropriate or yield biased evaluations when applied to newsroom data, which is typically noisy and often contains a significant number of outliers.
For example, the widely used Silhouette score, while effective for qualifying clustering quality on clean datasets, may not provide a fair or accurate assessment of the true clustering quality when handling typical newsroom data, such as reader comments, which can be more complicated and noisy. Ultimately, this can lead to AI-generated insights that are inevitably biased, undermining their credibility and trustworthiness.

- *Human and implementation bias:* Human choices in designing, training, and selecting AI models and algorithms or even vectorization models selection will also bring biases. For example, a newsroom data team, perhaps due to limited resources, consistently chooses simpler clustering algorithms (like basic K-Means) to analyze reader comments. They might also stick to a standard TF-IDF vectorization without exploring more advanced embeddings that could capture semantic nuance better. But K-Means' preference for spherical and equal-sized groups systematically overlooks smaller or irregular opinion groups, while TF-IDF systematically weights niche terms, burying signals from underrepresented voices. As a result, the newsroom's insights remain oversimplified and skewed, misrepresenting the true diversity of reader feedback [13].

Algorithm and Model Basies Can be More Challenging. While data biases can be addressed through better data governance and human biases through training and diverse oversight, model and algorithm bias presents a particularly entrenched challenge. These biases are often intrinsic to the mathematical foundations of the algorithms themselves. For instance, the assumptions made by standard clustering algorithms about data distributions or group characteristics are not easily altered without fundamentally redesigning the algorithm. Newsrooms typically utilize off-the-shelf or widely available AI tools, and therefore inherit these built-in model limitations. The development of new, inherently fairer, and context-sensitive algorithms is a long-term research endeavor.

Naive Newsroom Clustering. To emphasize the challenge posed by algorithm and model bias, let's consider a common scenario: naive newsroom clustering. Such an approach typically employs a straightforward clustering model (e.g., K-Means) to group reader comments, often relying primarily on keyword similarity derived from vectorization methods like TF-IDF or word embeddings (e.g., Word2Vec). This naive application inherently risks overlooking or misrepresenting nuanced discourse. Specifically, it tends to favor the identification of large, dominant opinion groups, potentially dismissing viewpoints expressed by a smaller number of commenters as 'noise' or merging them into broader, less accurate categories. Furthermore, the clustering outcomes can inadvertently reflect biases embedded within the chosen vectorization techniques themselves, such as those present in pre-trained word embeddings or the feature-weighting logic of TF-IDF.

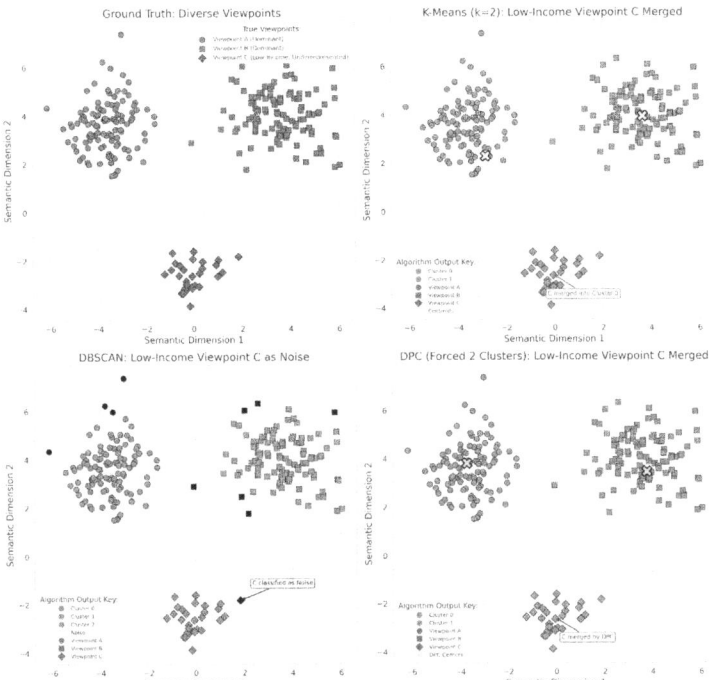

Fig. 2. Conceptual illustration of AI bias in comment clustering. Viewpoint C, representing a distinct minority viewpoint, might be incorrectly merged with dominant viewpoints by naive algorithms (e.g., K-Means, DPC) or labeled as noise (e.g., DBSCAN), leading to its marginalization in summarized insights

Figure 2 presents four scatter plots that reveal how naive clustering can obscure underrepresented viewpoints. In the top-left "Ground Truth" subplot, Viewpoint C stands out clearly as a small but distinct minority (low-income underrepresented)). In contrast, the top-right subplot shows K-Means (with k = 2) merging Viewpoint C into a larger cluster because it lacks sufficient clusters to separate it.

The bottom-left subplot illustrates DBSCAN labeling Viewpoint C as noise, while the bottom-right subplot forces DPC (Density Peak Clustering) into two clusters and again absorbs Viewpoint C into a dominant group. Together, these panels demonstrate how off-the-shelf algorithms can misclassify or ignore valid minority perspectives, a classic form of algorithmic bias. Unlike K-Means, which forces every point into a nearest-centroid sphere, DBSCAN clusters only where local point density exceeds a threshold, labeling sparse pockets as noise [14]. DPC selects points with the highest local density-and-distance as peaks and assigns neighbors to them [15]. Both capture irregular shapes but still erase low-density viewpoints—either by discarding them (DBSCAN) or merging them into dominant peaks (DPC).

This conceptual example is critical for journalism: if Group C corresponds to a real but less common reader perspective, its disappearance from the analysis prevents reporters from capturing the full range of audience reactions. Misrepresenting or omitting such views undermines the journalistic commitment to diversity and fairness. To uphold trust and ethical standards, clustering approaches must be carefully tuned to preserve minority voices rather than bury them.

3 Median-Based Silhouette Score: Fair and Explainable Clustering Measure

The widely-used silhouette score is not an explainable and fair measure for clustering newsroom data because it uses mean distances and can therefore be skewed by outliers and the noise prevalent in newsroom data. The noise or outliers range from readers' typos to vectorization noise. To achieve a more robust and ethically sound assessment, we introduce a median-based silhouette score.

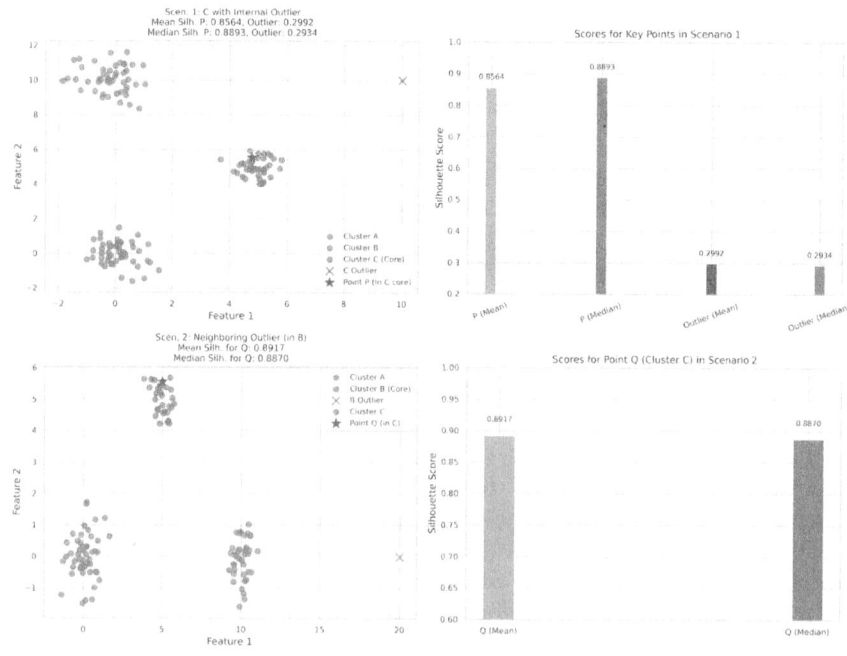

Fig. 3. Median-Based Silhouette demonstrates enhanced fairness and explainability. (Top row) An internal outlier within Cluster C disproportionately lowers the Mean Silhouette of a core point (P), while the Median Silhouette remains robust. (Bottom row) An outlier in a neighboring cluster (B) inflates the Mean Silhouette of a point in Cluster C (Q), whereas the Median Silhouette offers a more trustworthy measure of separation.

Median-Based Silhouette Scores. Let X be a dataset of N points, each $i \in X$ assigned to cluster C_i, and let $\text{dist}(i,j)$ denote the distance between points i and j. Define

$$a_{\text{med}}(i) = \text{median}_{j \in C_i \setminus \{i\}} [\text{dist}(i,j)] \tag{1}$$

$$b_{\text{med}}(i) = \min_{k \neq i} \text{median}_{j \in C_k} [\text{dist}(i,j)] \tag{2}$$

$$s_{\text{med}}(i) = \frac{b_{\text{med}}(i) - a_{\text{med}}(i)}{\max\{a_{\text{med}}(i), b_{\text{med}}(i)\}}, \tag{3}$$

Here $s_{\text{med}}(i) \in [-1, 1]$, where 1 means point i is very well clustered, 0 means it lies on a cluster boundary, and -1 means it's likely misclassified. Then S_{med} is the average over all N points:

$$S_{\text{med}} = \frac{1}{N} \sum_{i=1}^{N} s_{\text{med}}(i) \tag{4}$$

The Median-based silhouette score provides a more robust and fair evaluation of newsroom data clustering by minimizing the distorting impact of outliers and irregularly shaped opinion groups inherent in such noisy, real-world text.

Figure 3 shows two toy scenarios showing how the proposed median-based silhouette makes cluster quality scores fairer and more explainable in the presence of noisy or irregular data (outliers).

Scenario 1 (top row): an internal outlier within Cluster C disproportionately lowers the mean silhouette of a core point (P) to 0.856, while the median silhouette remains more robust at 0.889. For the outlier itself, both mean (0.2992) and median (0.2934) silhouette scores are low, correctly identifying its poor fit.

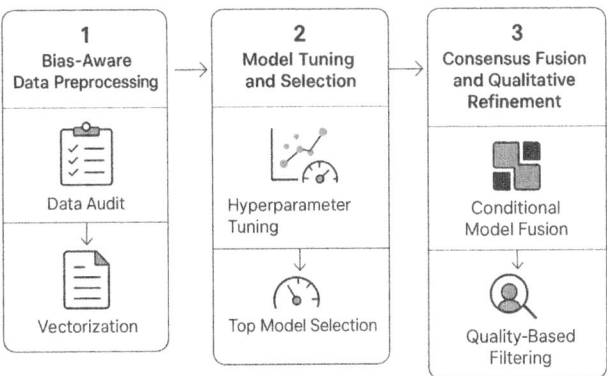

Fig. 4. The ethical newsroom clustering has three components: bias-aware preprocessing, model tuning & selection, and consensus fusion and quality filtering.

However, the median silhouette calculation for point P is less distorted by the outlier, offering a fairer assessment of P's good fit.

Scenario 2 (bottom row): A distant outlier of from a neighbor cluster B inflates the mean silhouette of the coe point Q of cluster C: 0.8917; But the median silhouette provides a correction, avoiding distortion by the stray point and reflecting true cluster distinctiveness.

It is noted that the global median silhouette score $S_{\text{med}} = \frac{1}{N}\sum_{i=1}^{N} s_{\text{med}}(i)$ in clustering will also be more trustworthy and robust to outliers than the existing global mean silhouette score because it is the average of the individual robust point scores; whereas, this does not necessarily mean the value of the global median silhouette will be higher numerically.

4 Ethical Newsroom Clustering: A Robust Framework

To mitigate AI-induced bias in the newsroom due to noisy inputs, simplistic algorithms, and human judgment errors, we propose *Ethical Newsroom Clustering*, a three-component pipeline that emphasizes transparent and ethical processing (see Fig. 4).

1. Bias-Aware Data Preprocessing

1. *Data Audit:* Audit data sources (e.g. comment archives) to identify potentially under- or overrepresented segments. If ethically sound, conduct re-weighting or data augmentation for minority segments.
2. *Bias-Aware Vectorization:* Select text vectorization methods (e.g., pre-trained embeddings, TF-IDF variants) to minimize inherent biases. This may involve checking embeddings for known societal biases (e.g., using WEAT [16]) or choosing TF-IDF weighting schemes that preserve important, less common terms crucial for representing diverse viewpoints.

2. Model Tuning and Selection

1. *Performance-Based Tuning:* We tuned each algorithm's hyperparameters for a robust median silhouette score, the preferred metric for t-SNE embeddings, because it filters vectorization noise and avoids extra computation. Metric gains guided, but never dictated, model selection; interpretability and practical relevance remained paramount.
2. *Top Model Selection:* From the tuned candidates, select the top two models based on their achieved median silhouette scores.

3. Consensus Fusion and Qualitative Refinement

1. *Conditional Model Fusion:* If the median silhouette score difference between the top two selected models is below a defined threshold (e.g., $\theta = 0.1$), conduct model fusion. Employ methods like a weighted Meta-Clustering Algorithm (MCLA) or similar ones to conduct consensus clustering and output the

model with the highest performance in terms of median silhouette scores [17]. Pick the top model if the consensus model has the same level performance.
2. *Quality-Based Filtering:* Define minimum quality cutoffs (e.g., $\tau_s < -0.5$) for the median silhouette score) for the final clusters. Clusters falling below such thresholds are flagged; their points may then be treated as noise or subjected to further qualitative human review to ensure that no significant, underrepresented viewpoints are inadvertently discarded and that reported groupings are trustworthy.

It is noted that we recommend customized models to start naive clustering before tuning. For example, K-Means for compact, spherical clusters; DBSCAN for arbitrary shapes and noise filtering; Affinity Propagation for exemplar-based grouping without fixing k [18]; and DPC (Density Peaks Clustering) for locating thematic peaks via local density and separation—and tune hyperparameters (e.g., k, ε, damping) to uncover minority viewpoints and prevent their absorption by dominant groups. It is not recommended to invite a clustering model that is not matched with the nature of input data, as this can lead to inherently flawed starting points. For example, directly applying a Gaussian Mixture Model (GMM) to raw, high-dimensional, and typically non-Gaussian text features (such as TF-IDF vectors from comments) would likely be inappropriate [19].

Furthermore, note that the median silhouette ignores the few extreme distances that can distort the mean silhouette, so we avoid hyperparameters that merely isolate noisy points. Because the median silhouette score smooths out the randomness of t-SNE's stochastic initialization, it yields steadier scores across runs and lets us narrow the hyperparameter search space.

In addition, unlike standard MCLA that typically builds a meta-graph using the simple overlap (shared points) between base clusters to determine their consensus, weighted MCLA enhances this by adjusting the strength of these relationships in the meta-graph (e.g., edge weights) based on how confidently each base model assigned the shared points to those clusters, often using individual-point median silhouette scores.

Therefore, weighted MCLA allows the final consensus to be more influenced by primary clustering model assignments with higher confidence, rather than treating all shared points equally. In our implemented version, weighting was done by multiplying each data point's one-hot encoded base cluster label with its corresponding scaled, individual point median-silhouette score from that base model before these were combined into meta-features, which are then new numerical representations for each data point reflecting its weighted membership across the different base clusterings, upon which a final meta-clustering algorithm is applied.

5 Experiments

To demonstrate the practical application and benefits of our Ethical Clustering framework, we conduct experiments on two datasets: (1) a synthetic dataset designed to contain known biases and distinct opinion groups, and (2) a real-world dataset of reader comments from The New York Times (NYT) articles. Our goal is to show how a naive clustering approach can lead to biased or incomplete interpretations, and how proposed ethical newsroom clustering can gain representative insights.

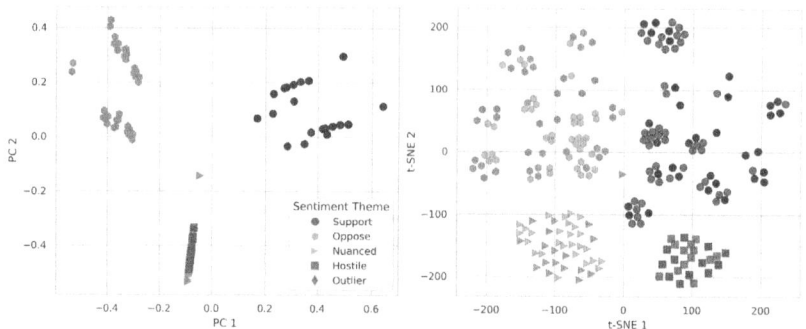

Fig. 5. Comparison of PCA and t-SNE visualizations for synthetic reader comment data. While PCA shows considerable overlap, t-SNE effectively separates the four primary sentiment themes ('Support', 'Oppose', 'Nuanced', and 'Hostile') into distinct visual clusters

5.1 Synthetic Reader Comment Data

Synthetic Dataset in Brief. The synthetic corpus contains 505 short "reader comments," distributed across four thematic classes—Support (200), Oppose (180), Nuanced (70), and Hostile (50)—plus a small Noise group and 5 deliberately crafted outliers. Support texts feature positive-policy phrases such as fantastic and fully support; Oppose comments cite economic objections (waste money, strongly oppose). Nuanced entries raise community-impact questions (e.g., "concerned about accessibility"), while Hostile ones mimic abusive or off-topic rants. This controlled yet imbalanced set, with known labels, gives a clean benchmark for evaluating how well clustering models recover the true classes, capture minority groups, and isolate noise—both individually and in consensus settings.

Tailored TF-IDF Vectorization. TF-IDF fits the synthetic comment data well: it amplifies the controlled, category-specific vocabulary by up-weighting distinctive keywords, down-weighting common tokens, and normalizing for the uniformly short comment lengths. Because these texts are brief and clean—free of typos, slang, emojis, or code-switching—the vocabulary remains compact, letting TF-IDF capture distances without fragmentation. By contrast, a heavyweight contextual model like BERT would introduce extra complexity without meaningfully improving separability over this simple, well-matched representation.

Figure 5 shows the PCA and t-SNE visualizations of this synthetic dataset under a TF-IDF representation, which aligns well with such synthetic comment data. Compared to PCA, t-SNE achieves better cluster separation, although all five outlier points remain embedded among the main clusters.

Involved Clustering Models: We used K-Means ($k = 4$), DPC (target $k = 4$, dc via 5th percentile of distances), and DBSCAN ($\epsilon = 1.25$, minPts=4). K-Means underperformed, likely because its spherical assumptions and forced outlier assignment distort cluster shapes, unlike the more flexible density-based DBSCAN (which also isolates noise) and DPC.

K-Means struggles because it assumes spherical, equally-sized clusters and forces all points (including outliers) into one of the k groups, distorting cluster shapes. DBSCAN excels by finding arbitrarily shaped clusters based on density and importantly, can identify and exclude outliers as noise. DPC also leverages density to find cluster centers, making it more robust to varied cluster shapes than K-Means, though simpler versions still assign all points.

Table 1. Sub-clusters within the Support class discovered by DPC.

Cluster ID	Size	% of Support	Typical tone (examples)
3	100	50%	Strong, enthusiastic—"fully support", "great idea"
2	44	22%	Measured positives—"makes sense", "positive change"
1	28	14%	Conditional support—"if budget allows"
0	28	14%	Pragmatic, cost-focused—"worth the expense", "good ROI"

In our ethical newsroom clustering, DPC, which is the final output model, delivers the strongest result: it forms seven clusters overall, giving each minority label its own pure group—Oppose (180), Nuanced (70), and Hostile (50)—and dividing the 200 Support comments into four clear sub-clusters. These are: (i) an upbeat core of 100 emphatic endorsements ("great idea"); (ii) 44 comments expressing measured positivity ("positive change"); (iii) 28 comments that offer conditional backing ("support if budget allows"); and (iv) 28 pragmatic remarks that support the proposal because the expected return on investment outweighs the cost (see Table 1).

Model Comparison. Table 2 reports the *median silhouette*—a robust cohesion/separation measure that down-weights extreme points—together with class-balanced scores for the three tuned models:

- *K-means* (k = 3) attains the highest *mean* silhouette (0.444), but with only three clusters it absorbs all 50 *Hostile* comments, driving its Macro-F1 (0.655) and Balanced Accuracy (0.722) to the bottom.

Table 2. Performance comparison of three clustering models on t-SNE embeddings.

Algorithm	Mean Silh	Median Silh	Clusters	Balanced d-index	Macro-F1	Balanced Acc.
K-means	0.444	0.436	3	1.817	0.655	0.722
DBSCAN	0.435	0.447	6	1.882	0.694	0.875
DPC	**0.421**	**0.499**	**7**	**1.884**	**0.859**	**0.893**

– DBSCAN ($\varepsilon = 47$, min_samples = 18) retrieves the Hostile class, yet its fixed-radius rule leaves 29 comments as noise and fragments the large *support* group, limiting its Macro-F1 to 0.694.
– DPC (target_k = 7, dc_percentile = 5) records the best median silhouette (0.499) and tops every fairness metric—balanced d-index 1.884, Macro-F1 0.859, Balanced Accuracy 0.893—because it assigns every sample to a cluster, keeps each minority label pure, and splits the 200 *Support* comments into four coherent sub-clusters [20,21].

Why the Median Silhouette Favours DPC. For DPC the median silhouette (0.499) is noticeably higher than its mean (0.421), indicating that most points sit well inside cohesive clusters while only a few outliers pull the mean down. This gap highlights DPC's strength: it delivers reliable cluster membership for the *typical* comment, not just on average, providing a more robust segmentation than the competing methods.

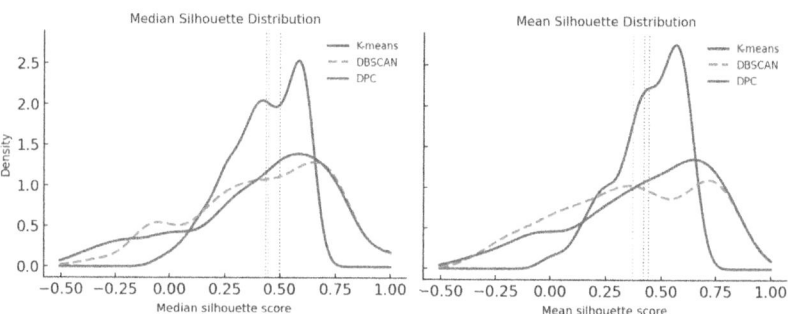

Fig. 6. Per-point silhouette distributions for the three tuned clustering models. (a) KDE (kernel density estimate) of median silhouette scores: DPC's curve sits farthest to the right, indicating the best typical cluster fit; dashed line marks DBSCAN.(b) KDE of mean silhouette scores, plotted on the same axis scale for direct comparison. Median-based curves (panel a) are preferred because they down-weight the few extreme points that can distort the mean, giving a more robust view of overall cluster quality.

Median and Mean Silhouette Distributions. Figure 6 compares per-point silhouette distributions. In the left plot, based on the median silhouette, a metric that weights outliers down, the density curve of DPC lies further to the right (median < 0.05), followed by DBSCAN (≈ 0.45) and K-means (≈ 0.44), showing that DPC offers the best group fit for the typical comment. The right plot uses the mean silhouette; because the mean is easily skewed by a few extreme points, the curves bunch closer and DBSCAN's long left tail is muted, illustrating why the mean is less reliable for hyper-parameter tuning and fairness assessment. Taken together, the figure underscores that median silhouette is the more robust metric and that DPC remains the top-performing model under this stricter lens.

Table 3. Performance when models tuned directly on the original TF-IDF data.

Algorithm	Mean Silh	Median Silh	Clusters	Noise
K-means ($k = 4$)	0.140	0.093	4	0
DBSCAN ($\varepsilon = 1.0$, $m_s = 5$)	**0.267**	0.116	8	0
DPC ($k = 4$, dc%=5)	0.196	**0.134**	4	0

Table 3 shows that DBSCAN attains the highest *mean* silhouette in the raw TF-IDF space, but its median drops to 0.116, indicating that only a few dense cores lift the average. DPC again shows a higher median (0.134) than mean (0.196), confirming its robustness for the *typical* comment, while K-means performs worst on both measures. Together with the stronger t-SNE results, these figures reinforce DPC's advantage across feature spaces.

5.2 Ethical Newsroom Clustering for New York Times Comments

Article and Reader Comments. We use the publicly available New York Times Reader Comments Kaggle corpus and focus on article ID: 5a4e7665 7c45 9f29 e79b 21da, titled "Trump Lawyers Seek to Block Fire and Fury Tell-All," published 18:45 UTC (1:45 p.m. ET) on 4 January 2018. The roughly 1,600-word story covers President Trump's bid—led by attorney Charles J. Harder—to halt publication of Michael Wolff's Fire and Fury: Inside the Trump White House by issuing cease-and-desist letters to Wolff, publisher Henry Holt, and former adviser Steve Bannon. The episode's First-Amendment stakes spurred strong reader engagement, generating 2,317 comments that serve as our analysis sample.

Challenge Handling in Data Preprocessing and Post Clustering. *Data Clean:* Such data often include redundant features that are unhelpful for clustering. We first removed non-essential fields—such as the article URL, userId, and userLocation, retaining only the comment text to prevent noise injection at the outset.

TF-IDF and BERT Falter. Additionaly, TF-IDF and vanilla BERT falter on such real comment data. TF-IDF may explode into a sparse, high-dimensional space where misspellings, slang, and emojis fragment the vocabulary and erase contextual meaning. On the other hand, vanilla BERT can output token features unsuited to distance-based clustering for such data. This is because real reader comments are short, slang-laden, and often sarcastic, so they split into many unequally important sub-word pieces; vanilla BERT gives each piece its own 768-d vector, and naïvely averaging those vectors mashes together typos and key stance words, yielding sentence embeddings whose pairwise distances no longer track the comments' true semantic or opinion differences.

RoBERTa Vectorization. We encode the cleaned comments with a RoBERTa based sentence transformer, an optimized version of BERT [22]; its sub-word tokenizer and large-scale pre-training retain nuance in misspellings, emojis, and slang, yielding compact sentence vectors that capture semantics far better than TF-IDF or vanilla BERT.

Post-clustering Analysis: Theme Inference: Unlike synthetic data, real reader-comment data lack ground-truth labels. It can also be hard to simply assume that comments are partitioned into support, oppose, hostile, neutral/nuanced, or other categories because it is possible that all comments are negative (oppose) but come from different perspectives and anchors.

To handle this, we conducted the following post-clustering analysis to infer theme for each cluster: 1) Surface the top five TF-IDF words for each cluster; proper nouns mark the topic, whereas evaluative verbs or adjectives indicate tone. 2) Review the ten comments closest to the TF-IDF centroid to verify that these cues dominate the discourse and to note the rhetorical style. 3) Run quick diagnostics: sentiment polarity with VADER and a named-entity count to see whether the focus is people, events, or documents [23]. 4) Combine topic and tone into a short label. If more than 20% of a random comment sample contradicts the draft label, refine it or re-examine the cluster.

Model Tuning and Selection. In addition to the density-peak clustering (DPC), centroid-based k-means, and density-based DBSCAN baselines, we apply spectral clustering to the t-SNE embedding of the RoBERTa vectors. Spectral clustering constructs a similarity graph and partitions it via the graph Laplacian, enabling recovery of non-convex, manifold-shaped clusters that centroid or density methods often miss [24]. It therefore complements DPC's density sensitivity and k-means' spherical bias by fully exploiting the local-neighbourhood structure revealed by t-SNE. The number of clusters k is chosen with the eigengap heuristic, and the final solution is selected using the median silhouette criterion.

Table 4 shows that spectral clustering delivers the strongest structure on the t-SNE map: it attains the highest mean and median silhouette scores (0.386/0.436) with only three clusters, indicating compact, well-separated groups. K-means reaches comparable cohesion but requires five clusters, while DBSCAN's low mean silhouette (0.162) reflects its tendency to label many boundary points as noise. DPC sits between these extremes (0.3173/0.4106). Although the weighted MCLA-based consensus clustering (spectral $\oplus$ K-means),

Table 4. Clustering scores of 4 models and a consensus model.

Model	Mean Sil.	Median Sil.	# Cluster	Key settings
K–Means	0.3817	0.4283	5	$k = 5$
DBSCAN	0.1618	0.3305	3	$\varepsilon = 4.75$, minPts $= 14$
DPC	0.3173	0.4106	4	target $k = 4$, $dc\_\% = 1$
Spectral	0.3859	0.4363	3	$k = 3$
Consensus	0.3859	0.4363	3	Spectral + K–Means

where spectral clustering is the primary model, reaches the same silhouette scores, we choose the spectral as the final model because it achieves the same level performance with less complexity according to our ethical clustering principle.

Table 5. Spectral clusters: keyword and inferred themes.

Cluster (#)	Keyword signals	Sentiment cues	Inferred theme
0 (847)	*trump, bannon, lost, mind, president*	Very negative polarity	Personal-derision
1 (838)	*kettle, pot, popcorn, people, trump*	Mildly negative; humour markers	Sarcastic cynicism/mockery
2 (632)	*book, mueller, fire, fury, house*	Neutral tone; dense proper nouns	Evidence-based critique/"Fire-and-Fury" focus

Table 5 lists the three spectral clusters identified by our post-clustering theme-inference pipeline: C0: Personal-derision (very negative: 847 comments), C1: Sarcastic cynicism (mildly negative:838 comments), and C2: Evidence-based critique (neutral: 632 comments). Together, these labels show that the comment stream is largely critical, albeit expressed in different angles.

Figure 7(a) further confirms this reading. Each cluster forms a dense, smoothly contoured region, and the kernel-density envelopes make the boundaries easy to see. The global median silhouette value of 0.436 confirms that the clusters are compact and well separated.

Figure 7(b) identified all 54 comments whose median-silhouette scores are below zero: 16 originate in cluster C0, 37 in C1, and only one in C2. Interestingly, these cases are not "bad data." Instead, they form a narrow ring of boundary points that separate the three clusters. Every score lies in the tight interval $[-0.1675, 0.0022]$, meaning each comment is almost equally similar to its own centroid and to the nearest neighbouring cluster. Rather than weakening the model, these near-zero values highlight where one rhetorical style shades into another.

The content of the boundary comments confirms this reading. Many C0→C1 points blend personal ridicule with popcorn-style sarcasm; C1→C2 points introduce evidence while retaining a wry tone; the lone C2 outlier invokes historical criticism (Nixon/Agnew) but omits Fire-and-Fury specifics. Keeping such comments is valuable: they delineate the discourse continuum, illustrate how themes overlap, and prevent the artificial hardening of cluster borders that can occur if borderline samples are discarded.

(a) Spectral clustering of reader comments (median silhouette = 0.436).

(b) Low-silhouette (< 0) boundary points from the initial solution.

Fig. 7. t-SNE visualization of the corpus after spectral clustering: (a) t-SNE view after spectral clustering: C0 Personal-derision (blue), C1 Sarcastic cynicism (orange), C2 Evidence-based critique (green), each bounded by KDE contours. (b) t-SNE with low median-silhouette points: red × (median-silhouette < 0) sit on the boundaries between C0 Personal-derision (blue), C1 Sarcastic cynicism (orange) and C2 Evidence-based critique (green), flagging borderline or ambiguous comments. (Color figure online)

6 Discussion

This paper has explored the critical ethical challenge of AI bias within newsroom applications, particularly in the analysis of unstructured text data like reader comments. To the best of our knowledge, it is the first study on this topic in AI journalism.

Our proposed ethical newsroom clustering framework demonstrated a practical pathway to mitigate such biases. The introduction of the median-based silhouette score proved crucial, offering a more robust and fair evaluation metric compared to the traditional mean silhouette, especially in the presence of outliers and imbalanced data characteristic of newsroom datasets. This was evident in its ability to guide hyperparameter tuning and model selection towards more equitable outcomes.

Experiments on synthetic data highlighted the ability of the proposed framework to preserve minority viewpoints, effectively identifying nuanced subgroups that naive approaches obscured. The application to New York Times comments further showed the framework's robust real-world utility. RoBERTa-based vectorization followed by spectral clustering, selected via our ethical pipeline, successfully delineated distinct, meaningful thematic clusters from a complex comment stream, revealing different angles of critique rather than a monolithic sentiment. Notably, the analysis of low-silhouette points as boundary cases rather than mere noise adds a layer of interpretability to clustering results, suggesting areas of thematic overlap rather than data deficiencies.

More broadly, the bias-aware preprocessing, fairness-centered metrics, and conditional fusion introduced in this framework can be readily extended to adjacent newsroom tasks: quote attribution, beat discovery, and even AI-assisted headline generation.

Complexity Does Not Always Buy Quality. The weighted MCLA consensus ensemble delivered the same median silhouette as the best standalone model (spectral clustering), but demanded an additional fusion step, two hyperparameter grids, and additional interpretability checks. In deadline-driven newsrooms, such marginal gains rarely justify the overhead. Future ensemble work should focus on directional advances, e.g., diversity-driven model selection or task-aware voting, rather than ever larger model pools.

A key limitation is the continued need for human oversight, especially when interpreting cluster themes and setting a sensible hyperparameter range. For instance, exploring configurations that produce large cluster counts (e.g., > 10) is seldom worthwhile, as such fine-grained partitions add little interpretive or analytical value in the newsroom.

Moreover, scalability to massive, real-time datasets also warrants further investigation. Future work should explore the adaptation of this ethical framework to other AI-driven journalistic tasks, such as automated story generation, bias detection in source selection, or fact-checking systems. Developing more automated methods for bias detection and mitigation within the preprocessing and model selection stages would also enhance its utility.

7 Conclusion

This study introduced ethical newsroom clustering, a novel framework designed to address and mitigate AI bias in the journalistic analysis of reader comments and similar unstructured text data. Our experiments, on both synthetic and real-world news comment data, validate the framework's efficacy in uncovering nuanced perspectives and bringing insights in the newsroom and avoid possible AI biases. The proposed median-based silhouette score can also be extended to enhance fairness and trustworthiness in journalism applications that involve unstructured data with noise.

As AI continues to integrate into newsrooms, such ethically-grounded frameworks are not merely beneficial but essential for upholding the core journalistic values of truth, accuracy, and fairness. This research offers a concrete methodology for newsrooms to harness AI's power responsibly, fostering analytical depth while safeguarding against algorithmic bias [25]. We believe that this work provides a valuable step towards ensuring that 'ethical bytes' become a foundational component of AI's future in journalism, ultimately strengthening public trust and the integrity of the information ecosystem besides bringing enrichment to ethical AI and its applications.

Acknowledgment. The first author expresses her deep gratitude to Dr. Henry Han for his mentoring and help with this project. We also appreciate anonymous reviewers for their insightful suggestions and comments that contributed to enhancing this article. This work is partially supported by McCollum endowed chair start-up fund.

References

1. Makhortykh, M., Helberger, N., Harambam, J., Bountouridis, D.: We are what we click: understanding time and content-based habits of online news readers. New Media Soc. **23**(9), 2773–2800 (2020)
2. Associated Press: Automated Insights for AP. (Illustrative: actual report or webpage on their automation efforts would be cited here) (2015)
3. Kim, D., Kim, S.: Newspaper companies' determinants in adopting robot journalism. Technol. Forecast. Soc. Chang. **117**, 184–195 (2017)
4. Lao, Y., You, Y.: Unraveling generative AI in BBC News: application, impact, literacy and governance. TGPPP (2024). https://doi.org/10.1108/TG-01-2024-0022
5. Chartbeat: Engagement Analytics. Relevant Chartbeat documentation or case study, Illustrative (2022)
6. NewsWhip: Real-time Media Monitoring. Relevant NewsWhip documentation or case study, Illustrative (2024)
7. Anagnostopoulou, A., Gouvea, T.S., Sonntag, D.: Enhancing journalism with AI: a study of contextualized image captioning for news articles using LLMs and LMMs, *arXiv preprint* arXiv:2408.04331 (2024)
8. Moravec, V., Hynek, N., Skare, M., Gavurova, B., Kubak, M.: Human or machine? The perception of artificial intelligence in journalism, its socio-economic conditions, and technological developments toward the digital future. Technol. Forecast. Soc. Chang. **200**, 123162 (2024)
9. Vosoughi, S., Roy, D., Aral, S.: The spread of true and false news online. Science **359**(6380), 1146–1151 (2018)
10. Bulger, M., Davison, P.: The promises, challenges, and futures of artificial intelligence and journalism. Digit. J. **6**(7), 807–818 (2018)
11. European Commission: Proposal for a Regulation of the European Parliament and of the Council laying down harmonized rules on artificial intelligence (Artificial Intelligence Act) (COM(2021)206 final) (2021)
12. Nascimento, F.R.S., Cavalcanti, G.D.C., Costa-Abreu, M.D.: Gender bias detection on hate speech classification: an analysis at feature-level. Neural Comput. Appl. **37**, 3887–3905 (2025)
13. Chierichetti, F., Kumar, R., Lattanzi, S., Vassilvitskii, S.: Fair clustering through fairlets. In: Guyon, I., Luxburg, U.V., Bengio, S., Wallach, H., Fergus, R., Vishwanathan, S., Garnett, R. (eds.) Advances in Neural Information Processing Systems, vol. 30, pp. 5029–5037. Curran Associates, Inc. (2017)
14. Ester, M., Kriegel, H.-P., Sander, J., Xu, X.: A density-based algorithm for discovering clusters in large spatial databases with noise. In: Proceedings of the 2nd International Conference on Knowledge Discovery and Data Mining (KDD), pp. 226–231 (1996)

15. Rodríguez, A., Laio, A.: Clustering by fast search and find of density peaks. Science **344**(6191), 1492–1496 (2014)
16. Caliskan, A., Bryson, J.J., Narayanan, A.: Semantics derived automatically from language corpora contain human-like biases. Science **356**(6334), 183–186 (2017)
17. Strehl, A., Ghosh, J.: Cluster ensembles – a knowledge reuse framework for combining multiple partitions. J. Mach. Learn. Res. **3**, 583–617 (2003)
18. Frey, B.J., Dueck, D.: Clustering by passing messages between data points. Science **315**(5814), 972–976 (2007)
19. Banerjee, A., Dhillon, I.S., Ghosh, J., Sra, S.: Clustering on the unit hypersphere using von Mises-Fisher distributions. J. Mach. Learn. Res. **6**, 1345–1382 (2005)
20. Han, H., Wu, Y., Wang, J., Han, A.: Interpretable machine learning assessment. Neurocomputing **561** (2023). Article 126891
21. Han, H., Li, D., Liu, W., Zhang, H., Wang, J.: High dimensional mislabeled learning. Neurocomputing **573** (2024). Article 127218
22. Reimers, N., Gurevych, I.: Sentence-BERT: sentence embeddings using Siamese BERT-Networks. In: Proceedings of the EMNLP–IJCNLP, pp. 3982–3992 (2019)
23. Hutto, C.J., Gilbert, E.: VADER: a parsimonious rule-based model for sentiment analysis of social media text. In: Proceedings of the 8th International AAAI Conference on Weblogs and Social Media (ICWSM), pp. 216–225 (2014)
24. Ng, A.Y., Jordan, M.I., Weiss, Y.: On spectral clustering: analysis and an algorithm. In: Dietterich, T.G., Becker, S., Ghahramani, Z. (eds.) Advances in Neural Information Processing Systems, vol. 14, pp. 849–856. MIT Press (2002)
25. Han, H.: Challenges of reproducible AI in biomedical data science. BMC Med. Genom. 18(Suppl. 1) (2025). Article 8

New Data-Science Infrastructure and Platforms

Unleashing Mojo: Accelerating K-Nearest Neighbor Learning

Sumanth Kolli, Chujiang Wu, and Henry Han[✉]

Data Science and Artificial Intelligence Innovation Laboratory, School of Engineering and Computer Science, Baylor University, Waco, TX 76798, USA
Henry_Han@baylor.edu

Abstract. New compiled languages such as *Mojo*, equipped with native SIMD kernels and explicit thread-level parallelism, promise to raise the performance ceiling that pure-Python machine learning (ML) pipelines often hit. We therefore re-implemented brute-force k-nearest neighbours (k-NN) in Mojo—combining 64-byte-aligned buffers, vectorized Euclidean kernels and lock-free thread pools—and benchmarked it against *scikit-learn*'s canonical Python/Cython implementation. The evaluation spans eleven datasets that vary along three orthogonal axes: sample count (10^2–10^6), dimensionality (4–3 072) and structural regularity (MNIST, CIFAR and synthetic blobs). Each experiment was repeated on a laptop-class six-core CPU and a workstation-class sixteen-core CPU to expose hardware effects. Mojo accelerates structured, cache-friendly workloads by five- to ninety-fold and sustains speed-ups of up to 60% on medium-scale image sets even in single-core mode. The margin shrinks to at most 10% on million-point or ultra-wide tables and can invert on low-end laptops when memory bandwidth dominates. These findings show that Mojo is a convenient accelerator for medium-sized, latency-sensitive applications, e.g. recommendation engines and edge analytics—whereas *scikit-learn* remains the pragmatic choice once datasets outgrow on-chip resources or when ecosystem maturity outweighs raw speed. All code, timing logs, and hardware counters are available in an open repository to facilitate reproduction and further optimization.

Keywords: KNN · Mojo · SIMD · Benchmarking · scikit-learn

1 Introduction

The k-Nearest Neighbors (k-NN) algorithm remains a fundamental component of many machine learning (ML) systems due to its intuitive design and effectiveness in both classification and regression contexts [1]. As a non-parametric method, it makes no prior assumptions about data distribution, which contributes to its broad applicability. However, the algorithm's reliance on exhaustive distance computations between query samples and all training instances results in considerable computational overhead, particularly when applied to high-dimensional or large-scale datasets even with speedup techniques like k-d trees [2]. These performance limitations constrain k-NN's deployment in latency-sensitive or resource-constrained environments, where algorithmic efficiency is critical.

In practice, one of the most accessible and widely adopted implementations of k-NN is provided by *scikit-learn* (sklearn), a Python-based ML library that prioritizes ease of use and broad functionality, while incorporating C-based methods for algorithmic efficiency [3]. For example, Sklearn's `KNeighborsClassifier` offers a standardized API for applying k-NN to a variety of tasks, leveraging efficient Cython back-ends for distance calculations and support for parallel processing via multi-threading. Despite these optimizations, sklearn remains constrained by the performance ceilings of the Python ecosystem and the inherent costs of run-time interpretation, particularly when scaling to millions of data points or integrating with hardware-level acceleration. As such, while sklearn's k-NN provides an excellent pedagogical and prototyping tool, its suitability for performance-critical production applications can be limited.

To mitigate these limitations, recent developments in programming language design have introduced tools optimized for high-performance computing and intensive AI workloads. One such language, Mojo, is a compiled language that combines strong static typing with explicit memory management, SIMD (Single Instruction, Multiple Data) vectorization, and built-in support for parallel execution - while maintaining seamless interoperability with existing Python code and libraries [6].

These capabilities position Mojo as a potentially advantageous alternative to traditional interpreted environments like Python, especially in workloads dominated by low-level numerical operations. Mojo has been widely advertised as significantly outperforming Python in specific benchmarks, with some promotional materials claiming execution speeds up to 68 000 times faster in numerical-computing tasks [5]. These claims, while impressive, stem primarily from controlled benchmarks and promotional content, and thus warrant further empirical evaluation in varied real-world contexts.

This study conducts a comparative analysis of k-NN implementations in Mojo and scikit-learn to assess the practical impact of the performance-oriented features of Mojo. The evaluation is guided by the following key questions.

- Under what dataset conditions, such as varying size, structural regularity, and dimensionality, does Mojo demonstrate tangible runtime improvements?
- To what extent do Mojo's low-level optimizations translate into meaningful advantages for performance-critical ML tasks?

Addressing these questions is pivotal for AI engineers weighing whether Mojo's promise justifies migrating away from the deep, battle-tested Python stack in latency-critical inference pipelines and edge deployments. Equally important, mapping Mojo's low-level optimizations to measurable end-to-end gains gives system architects clear guidance on where to focus engineering effort to maximize throughput and hardware efficiency in real-world ML workloads.

By exploring these questions, this study aims to provide clear evidence on Mojo's real-world applicability, thus offering valuable insights into optimal language and implementation selection for high-performance ML workflows. We have the following threefold contributions:

1. We present the first open-source, SIMD-vectorized, multithreaded implementation of brute-force k-NN in Mojo and analyze the design decisions—memory alignment, prefetching and lock-free work sharing—that unlock its speed. This work provides a practical blueprint for leveraging Mojo's low-level features to accelerate fundamental ML algorithms.
2. We perform an extensive benchmark on 11 datasets that span five orders of magnitude in sample size and three orders in feature dimensionality, executing on both laptop-class (6-core) and workstation-class (16-core) CPUs. These empirical results offer clear guidance on the real-world performance benefits and scalability of Mojo for KNN across diverse hardware and data regimes.
3. We release a complete, easy-to-reproduce artifact—including Mojo/Python source code, fixed train-test splits, benchmark scripts, timing logs, and raw hardware-counter traces—so that the community can independently verify and extend our evaluation of compiled languages on classical ML workloads.

This paper is organized as follows. Section 2 describes benchmark datasets, experimental setup, and detailed Mojo-k-NN implementation techniques; Sect. 3 presents and analyzes the results; Sect. 4 discusses implications, limitations, and directions for future work; and Sect. 5 concludes the article.

2 Method

This section details the experimental methodology employed for our comparative analysis. We describe the datasets and computational environments, outline the specifics of our Mojo k-NN implementation alongside the scikit-learn baseline, and detail the benchmarking protocol.

2.1 Datasets

We include 11 benchmark datasets, selected to span a wide range of characteristics relevant to k-NN performance: sample count (from 10^2 to over 10^5), feature dimensionality (from 4 to over 3,000), and data modality (numerical, image). These include standard benchmark datasets from the UCI Machine Learning Repository and scikit-learn's built-in datasets, widely recognized image classification datasets (MNIST, Fashion-MNIST, CIFAR-10), and synthetically generated datasets [7–10]. This breadth of scales and data types enables a rigorous examination of how Mojo's k-NN implementation copes with everything from small, low-dimensional "toy" problems to large, high-dimensional image workloads.

Synthetic data was created using scikit-learn utilities (*make_classification*, *make_blobs*) with fixed random seeds to ensure reproducibility and to test specific performance aspects under controlled conditions. For each dataset, a predefined train-test split was used consistently. The key characteristics of these data sets are summarized in the following Table.

Table 1 summarizes the 11 datasets used in our benchmark suite. They span four orders of magnitude in sample count, cover both low- and high-dimensional feature spaces, and include *vision, tabular,* and *synthetic* data. All public corpora were obtained through either the sklearn.datasets module or their original authors' websites; synthetic sets were generated with *scikit-learn* utilities (*make_classification, make_blobs*) using fixed random seeds to guarantee reproducibility.

Table 1. Datasets ($n \times d$ = training instances × features).

Dataset	$n \times d$	Test	Mod.	Source/reference
Public corpora				
Iris	105×4	45	Num.	UCI repository
Wine	143×13	35	Num.	UCI repository
Breast Cancer	455×30	114	Num.	UCI repository
Digits (8×8)	$1\,697 \times 64$	100	Img.	scikit-learn
MNIST (28×28)	$69\,895 \times 784$	105	Img.	LeCun et al. [8]
Fashion-MNIST	$69\,895 \times 784$	105	Img.	Xiao et al. [9]
CIFAR-10[†]	$55\,000 \times 3\,072$	5\,000	Img.	Krizhevsky et al. [10]
Covertype	$522\,910 \times 54$	58\,102	Num.	UCI repository
Synthetic datasets				
Synthetic Blob	$250\,000 \times 400$	1\,000	Num.	scikit-learn
Synthetic Uniform	$250\,000 \times 400$	1\,000	Num.	Uniform random generator
Synthetic High-Dim	$100\,000 \times 1\,000$	1\,000	Num.	scikit-learn

[†] RGB images flattened to a 3 072-element vector ($32 \times 32 \times 3$).

2.2 Experimental Setup

Hardware and Software Configuration. All benchmarks were conducted on two distinct systems to evaluate performance across different hardware classes:

- *Low-End Machine:* A laptop equipped with a 6-core/12-thread 2.6 GHz CPU and 16 GB of RAM.
- *High-End Machine:* A desktop workstation featuring an 8-core/16-thread 4.2 GHz CPU and 32 GB of RAM.

Both machines executed the benchmarks inside identical Ubuntu 24.04.2 environments running under WSL 2. Keeping the OS, kernel, and library versions fixed while varying only the CPU class (6-core laptop vs. 16-core workstation) isolates the impact of Mojo's low-level optimizations from confounding software stack effects. Pilot runs showed no measurable difference between WSL 2 and a native Linux boot on these CPU-bound workloads. The toolchain consisted of Mojo 25.1.1 and Python 3.12 with scikit-learn 1.5.0 (latest stable release at the time of testing).

2.3 Mojo k-NN

Our Mojo implementation pushes brute-force k-NN to the metal: training rows are 64-byte aligned and scanned with explicit SIMD kernels, while a lock-free thread pool processes query batches in parallel. Combined with cache-friendly prefetching, vector-register top k selection, and zero-copy data moves, these choices attack the algorithm's two bottlenecks—distance accumulation and neighbour retrieval—at both the data-layout and execution-model levels.

We then compare our Mojo k-NN implementation against the canonical brute-force k-NN classifier from scikit-learn. Both implementations compute exact nearest neighbors without approximate search structures (e.g., k-d trees or ball trees) to ensure a direct comparison of raw computational efficiency for the distance calculation and neighbor search phases.

Mojo k-NN Implementation. Our Mojo k-NN classifier (Algorithm 1) is designed for high performance by leveraging several low-level optimization techniques native to Mojo:

- *Memory layout and alignment:* The training matrix is held in one contiguous block and every row is padded to a 64-byte boundary so that each SIMD load falls within a single cache line, eliminating split-load penalties and maximising memory bandwidth [11]. Distances and neighbour indices are written to pre-allocated scratch buffers that are reused across queries, avoiding per-query allocations and keeping runtimes consistent.
- *SIMD-vectorized distance calculation:* Euclidean distances are computed with Mojo's explicit vector types (`vector<Float32, SIMD_WIDTH>`). The loop subtracts query chunks from aligned training chunks, squares the differences with fused multiply-add, and collapses the SIMD accumulator with one horizontal sum [12]. Two-line `__builtin_prefetch` hints keep the next training row in cache, mitigating memory on large matrices.
- *Efficient top-k selection:* A two-pronged strategy is used for identifying the k nearest neighbors. For small k (e.g., $k \leq 16$, a common scenario in many applications), a specialized SIMD-accelerated quicksort variant operating directly on vector registers is employed. This method minimizes memory access and leverages data-level parallelism for sorting small arrays. For larger k, a vectorized 'argpartition' equivalent is used to efficiently find the k smallest distances (and their indices) without incurring the cost of a full sort.
- *Lock-free multithreading:* A fixed-size thread pool is initialised once; each worker pulls query batches from a lock-free queue, computes distances to the shared (read-only) training set, and returns results without synchronisation barriers—eliminating lock contention and keeping every CPU core fully utilized.
- *Optimized data movement:* For internal operations such as preparing data rows for processing or consolidating results, low-level memory copy functions (e.g. `memcpy`) are utilized for fast block-wise data movement, which synergizes well with aligned memory layout [13].

The following algorithm summarizes our Mojo k-NN implementation—an SMB-k-NN (SIMD-vectorized, Multithreaded, Brute-force k-NN). Algorithm 1 (SMB-k-NN) runs in two stages.

1) *Distance phase* (lines 6–14): each thread scans the entire training matrix row by row, computes squared Euclidean distances in SIMD chunks (load64 + fused multiply-add), and writes the results into a thread-local buffer.
2) *Selection phase* (lines 15–20): depending on k, the thread uses either a register-level SIMD quicksort (for small k) or a vectorized partial partition, then performs a majority vote over the k nearest labels. The thread pool is fixed-size and lock-free: query vectors are statically partitioned across threads, so no synchronization is required once computation begins. All data structures that might cause contention (dist[], idx[]) are thread-local, yielding near-linear scaling until memory bandwidth becomes the bottleneck.

Algorithm 1. SIMD-vectorized, Multithreaded Brute-force k-NN(SMB-k-NN)

Input:
 Training set: $X \in \mathbb{R}^{n \times d}$ (rows 64-byte aligned), $\mathbf{y} \in \{1, \ldots, C\}^n$
 Query set: $Q \in \mathbb{R}^{m \times d}$
 neighbour count k
Output: $\hat{\mathbf{y}} \in \{1, \ldots, C\}^m$
 Constants: SIMD width V_W, tiny-k cutoff K_{tiny}
 Thread-local: $dist[n]$, $idx[n]$

1: **for all** threads t in pool **do** ▷ lock-free parallelism
2: **for all** query **q** owned by t **do**
3:   **for** $j \leftarrow 1$ to n **do** ▷ ℓ_2 to X_j
4:    $\mathbf{s} \leftarrow \mathbf{0}$ ▷ SIMD accumulator
5:    **for** $\ell \leftarrow 0$ to $d-1$ step V_W **do**
6:     $\mathbf{r} \leftarrow \text{load64}(X_{j,\ell:\ell+V_W-1}) - \mathbf{q}_{\ell:\ell+V_W-1}$
7:     $\mathbf{s} \leftarrow \mathbf{s} + (\mathbf{r} \odot \mathbf{r})$
8:    **end for**
9:    $dist[j] \leftarrow \text{sum}(\mathbf{s})$
10:   **end for**
11:   **if** $k \leq K_{\text{tiny}}$ **then**
12:    $idx \leftarrow \text{simdSort}(dist, k)$
13:   **else**
14:    $idx \leftarrow \text{argPartition}(dist, k)$
15:   **end if**
16:   $\hat{y}_{\mathbf{q}} \leftarrow \text{mode}(\mathbf{y}[idx])$
17: **end for**
18: **end for**
19: **return** $\hat{\mathbf{y}}$

Scikit-Learn k-NN Baseline. We employ KNeighborsClassifier in scikit-learn, with algorithm=brute for comparison, which performs the same row-wise, exact distance scan as our Mojo kernel [3]. Key implementation details are:

- **Compute kernel.** Distance loops are implemented in Cython and NumPy; when NumPy is linked against an optimized BLAS library (OpenBLAS, Intel MKL, etc.) the inner products are off-loaded automatically.
- **Top-k selection.** Neighbors are chosen with `std::nth_element` from the C++ STL, which runs in $O(n)$ average time [14].
- **Memory layout.** Both training and query matrices reside in standard C-order NumPy arrays [15].
- **Parallelism.** OpenMP is activated when `n_jobs`$\neq$ 1 (e.g., `-1` for all cores). Because the heavy lifting occurs inside C/Cython, these worker threads bypass Python's GIL, though minor GIL hand-offs can still occur during ancillary Python callbacks.

This setup is the fastest "out-of-the-box" exact k-NN available in the Python ecosystem and therefore serves as a strong reference point for judging Mojo's low-level optimizations.

Key Implementation Differences Between Moko k-NN and Sklearn k-NN. The primary distinctions between our Mojo implementation and the scikit-learn brute-force baseline are summarized in Table 2. These differences underscore Mojo's paradigm of enabling explicit, fine-grained control over memory and execution, contrasted with scikit-learn's approach of leveraging higher-level abstractions and established library-mediated optimizations.

Table 2. Comparison of Mojo and scikit-learn KNN implementations

Feature	Mojo Implementation	scikit-learn Implementation
Memory Layout	Manually aligned flat buffers with 64-byte alignment	NumPy arrays (C-order) with runtime-managed memory [16]
SIMD Utilization	Explicit vector types with hand-tuned intrinsics	Implicit, via Cython and BLAS
Top-k Selection	SIMD quick-sort and vectorized argpartition	Partial sort via C++ STL (`nth_element`)
Multithreading	Manual thread pool, pinned to CPU cores	OpenMP, subject to Python GIL limitations
Memory Management	Static allocation, zero runtime fragmentation	Dynamic allocation via Python runtime
Data Movement	`memcpy` used for aligned block copying	Internal copying via NumPy and Python buffers

2.4 Benchmarking Protocol

To ensure a robust and fair performance evaluation, the following protocol was meticulously adhered to:

- *Performance metric:* The primary metric for comparison was the wall-clock execution time consumed by the k-NN prediction phase on the designated test set. Time taken for initial data loading and one-time setup (such as the initial construction of Mojo's aligned matrix from raw data) was excluded from the timed measurements to isolate the inference speed.
- *Repetitions and averaging:* Each experimental run (defined as a specific dataset processed by a specific implementation on a particular hardware configuration) was repeated 10 times. The average execution time and the standard error of the mean (SEM) are reported to account for the variability of the measurement.
- *Neighborhood size (k):* Unless explicitly stated otherwise, a value of $k = 5$ was used for all k-NN classifications. This is a commonly used default value in practice.
- *Isolated Euclidean sistance benchmark:* For the specific benchmark focusing on the raw computational efficiency of Euclidean distance calculation (detailed in Sect. 3.1), all implementations (Pure Python, NumPy, Mojo) were constrained to execute on a single CPU core. This setup eliminates parallelization overheads and allows for a direct assessment of scalar and SIMD computational performance.
- *Full k-NN benchmark:* For end-to-end tests, both Mojo and scikit-learn used all available CPU cores (e.g. `n_jobs=-1`).
- *Statistical test:* Where performance comparisons are presented (e.g., in Table 4 and Table 5), p-values were computed using a two-sample Welch's t-test (assuming unequal variances) to assess the statistical significance of observed differences in mean runtimes between the Mojo and Python/scikit-learn implementations for each dataset.

All code, dataset scripts, timing logs and raw hardware counter traces are publicly available (see Sect.6) to enable full reproducibility and further optimization.

3 Results

3.1 Euclidean Distance Calculations

We compare the performance of using Mojo, Numpy, and Python in calculating pairwise Euclidean distances for two 5000 × 128 matrices randomly generated. This microbenchmark serves as a minimal but representative kernel to gauge the raw arithmetic and memory handling efficiency that ultimately limits end-to-end k-NN throughput because if the importance of distance calculation in this algorithm.

To ensure a fair and focused evaluation of raw computational performance, all implementations (Mojo, Numpy, and Python) were restricted to single-core, non-parallel execution. This isolates the efficiency of the underlying compiled code, allowing a clear comparison of how each environment handles basic floating-point operations.

Figure 1 compares their running time to calculate all pairwise Euclidean distances between the two 5000 × 128 matrices under Mojo, Numpy and Python, suggesting the leading performance of Mojo. The advantage of Mojo comes from the compilation ahead of time that exposes explicit SIMD vectorization and eliminates Python layer overhead, enabling it to saturate CPU floating point units far more effectively than NumPy's generalized BLAS calls or pure Python loops [4,6].

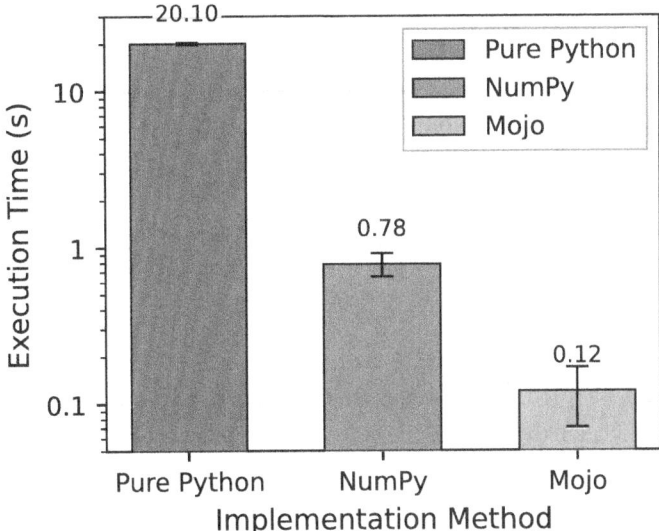

Fig. 1. Wall-clock time for computing all pairwise Euclidean distances between two 5000 × 128 matrices (single-core, no parallelism) under Mojo, Numpy, and Python. Bars show mean of 10 runs; whiskers indicate the standard error of the mean. Mojo's SIMD kernel shortens the calculation to 0.12 s, roughly 25× faster than NumPy's vectorized routine and 170× faster than naïve pure-Python code.

Table 3 shows that Mojo is about 167× faster than pure Python and more than 25× faster than NumPy under these conditions. It also demonstrated the best stability for its smallest SEM values compared to those of Numpy and python. Because Mojo's fully compiled kernel avoids interpreter jitter and dynamic allocator overhead, its execution path is highly deterministic, yielding the smallest standard-error-of-the-mean (SEM) among the three implementations [17]. These results foreshadow the overall superiority of our Mojo-based k-NN implementation, because distance calculation is a core component in k-NN.

Table 3. Euclidean-distance micro-benchmark.

Implementation	Running time ± SEM (s)	Relative speed
Pure Python	20.10 ± 0.30	1.00×
NumPy	0.782 ± 0.132	25.7×
Mojo	0.120 ± 0.050	167.5×

3.2 k-NN Performance Evaluation

The performance of our Mojo k-NN implementation relative to scikit-learn's brute-force classifier was evaluated across 11 datasets on both low-end and high-end hardware configurations. Benchmarking the same workloads on a commodity laptop as well as a workstation-class CPU lets us span the typical hardware spectrum—from resource-constrained edge devices to server-grade systems—so we can see whether Mojo's speed-ups hold consistently, shrink, or invert as compute resources, cache sizes, and memory bandwidth change. It is noted that the CIFAR and Cover datasets could not be run on a low-end laptop due to hardware limitations. The Cover data set also did not run using Mojo on a high-end PC.

Table 4. Comparison of Python vs Mojo runtimes on Low-End PC

Dataset	Python (s)	Mojo (s)	Speedup (%)	p-value
Iris	1.105 ± 0.141	0.179 ± 0.010	83.8%	<0.0001
Wine	0.013 ± 0.004	0.0003 ± 0.00002	97.8%	<0.0001
Cancer	0.020 ± 0.006	0.0005 ± 0.00003	97.4%	<0.0001
Digits	0.030 ± 0.005	0.0022 ± 0.00003	92.9%	<0.0001
MNIST	2.164 ± 0.096	1.252 ± 0.143	42.1%	<0.0001
Fashion	2.122 ± 0.060	1.216 ± 0.160	42.7%	<0.0001
Blob	32.916 ± 0.492	33.298 ± 1.573	−1.2%	0.47
Random	31.926 ± 0.443	33.296 ± 1.690	−4.3%	0.02
CIFAR	N/A	N/A	N/A	N/A
Cover	N/A	N/A	N/A	N/A

The results, summarized in Table 4 (Low-End PC) and Table 5 (High-End PC), and visualized in Fig. 2, reveal distinct performance characteristics contingent on the size, dimensionality and available computational resources of the data set. We have following detailed findings of Mojo's performance on small, median, large, and high-dimensional datasets.

Table 5. Comparison of Python vs Mojo runtimes on High-End PC

Dataset	Python (s)	Mojo (s)	Speedup (%)	p-value
Iris	0.011 ± 0.001	0.0002 ± 0.00005	98.1%	<0.0001
Wine	0.011 ± 0.001	0.0002 ± 0.00009	97.9%	<0.0001
Cancer	0.012 ± 0.001	0.0010 ± 0.0014	91.8%	<0.0001
Digits	0.011 ± 0.000	0.0009 ± 0.0001	91.8%	<0.0001
MNIST	0.541 ± 0.020	0.230 ± 0.011	57.5%	<0.0001
Fashion	0.520 ± 0.004	0.223 ± 0.019	57.1%	<0.0001
Blob	8.620 ± 0.312	8.140 ± 0.084	5.6%	0.0002
Random	8.653 ± 0.294	8.156 ± 0.050	5.7%	0.0001
CIFAR	76.841 ± 1.675	45.500 ± 0.186	40.8%	<0.0001
Cover	192.524 ± 0.917	N/A	N/A	N/A

Performance on Small-Sized Data. A consistent trend observed across both hardware platforms is the substantial acceleration achieved by Mojo on smaller and well-structured data sets. For instance, on datasets like Iris, Wine, Cancer, and Digits, Mojo demonstrates speedups ranging from approximately 90% to over 98% (e.g., Iris on High-End PC: Mojo 0.0002 s vs. Python 0.011 s). This pronounced advantage can be attributed to Mojo's significantly lower overhead compared to the Python/scikit-learn stack. For small data volumes, the fixed costs associated with Python's runtime interpretation, function call dispatch, and NumPy C-API interactions in scikit-learn can dominate the overall execution time. Mojo, as a compiled language with direct memory access and minimal runtime overhead, excels in these scenarios by efficiently executing the core computational loops.

Performance on Medium-Sized Data. On medium-sized data such as MNIST and Fashion-MNIST (each with approximately 70,000 samples and 784 features), Mojo maintains a robust performance lead, achieving speedups of around 42% on the low-end PC and approximately 57% on the high-end PC. These datasets are large enough for the computational workload to become significant, yet still benefit from Mojo's explicit SIMD vectorization and efficient memory management, which likely leads to better cache utilization and faster distance calculations compared to scikit-learn's more general-purpose optimizations.

Performance on Large-Scale Data. The performance landscape becomes more nuanced for large-scale datasets(e.g., Blob and Random, with 250,000 samples and 400 features). On the low-end laptop, Mojo exhibited a slight performance degradation (1–4% slowdown) for these datasets. This suggests that on resource-constrained hardware, factors like memory bandwidth limitations or less efficient cache hierarchies might negate some of Mojo's computational advantages, particularly if scikit-learn's underlying BLAS libraries are highly optimized for the specific CPU architecture. However, on the high-end workstation, Mojo regained a modest speedup of approximately 5–6% for these same datasets. This reversal highlights the importance of sufficient hardware resources (faster CPU, more memory bandwidth, larger caches) for Mojo's low-level optimizations to fully translate into end-to-end performance gains on larger data. Notably, on the high-end PC, Mojo consistently outperformed Python for all datasets that could be successfully executed by both implementations.

Performance on Data with Very High Dimensionality. For datasets characterized by very high dimensionality, such as CIFAR-10 (55,000 samples × 3,072 features), Mojo achieved a significant speedup of around 40% on the high-end PC. This outcome underscores the efficacy of Mojo's explicit SIMD vectorization in handling wide feature vectors, where parallel processing of feature dimensions during distance computation becomes critical. The ability to manually tune SIMD operations for specific vector widths allows Mojo to maximize throughput on such computationally intensive tasks [18].

Overall Performance Profile and Attributing Factors. In aggregate, the benchmark shows a clear interaction between dataset scale, feature width, and hardware resources. Mojo's largest gains appear where per-sample arithmetic dominates (small and medium sets, or very wide vectors), because its ahead-of-time compilation eliminates Python call overhead and exposes fully-vectorized inner loops that keep the FMA pipelines busy.

As sample counts and distance evaluations grow into the hundreds of thousands, memory-system effects begin to dominate: on the laptop-class CPU Mojo's compute units frequently stall waiting for data, eroding its advantage over scikit-learn's BLAS-backed kernel, whereas the workstation's larger caches and higher memory bandwidth restore a modest edge. In contrast, when the dimension explodes (CIFAR-10), the problem is once again compute bound: Each cache line provides many useful features, and Mojo's hand-tuned SIMD instructions translate into a 40 % win from end to end.

Together these results suggest that Mojo is most beneficial for workloads that (i) fit comfortably in LLC or (ii) possess extremely wide feature vectors, while Python/Cython implementations remain competitive once the working set saturates the memory hierarchy on constrained hardware.

Figure 2 condenses these findings into a single bar chart that plots the Mojo-over-Python speed-up for every dataset on both machines, immediately revealing where gains spike or vanish; by contrast, Fig. 3 breaks the data out into per-dataset runtime bars, letting the reader see absolute timings and error bars for each hardware tier side by side.

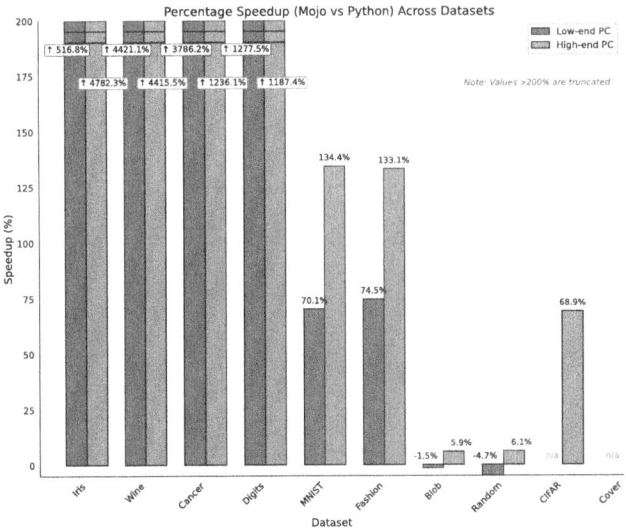

Fig. 2. Speedup comparison between Python and Mojo across different datasets and different PCs. Percentages above 200% are truncated but still represented. Values not present in both low-end PCs and high-end PCs are represented as N/A.

3.3 Empirical Runtime Complexity

We evaluated the empirical run-time complexity of Mojo k-NN by fitting a log-log regression to observed run-times across datasets of varying training sizes in Fig. 4. Using the model $\log T(n) = k \log n + \log C$, we estimated the effective complexity as $T(n) \approx C \cdot n^k$, where k reflects scaling behavior and C represents the constant baseline runtime cost.

The constant C captures hardware-independent efficiency, such as memory layout, data access speed, and language-level overhead. A smaller C indicates better low-level optimization. This is why C matters: while k tells us how fast runtime grows, C tells us how fast it starts. Understanding both is essential to choose the right implementation in production environments.

In the high-end system, we found $k \approx 1.33$ with $C \approx 2.07 \times 10^{-7}$; in the low-end system, $k \approx 1.19$ with $C \approx 3.80 \times 10^{-6}$. Although the high-end PC delivers faster runtime, its higher complexity exponent suggests steeper growth as data size increases. In contrast, the low-end PC scales more gradually but incurs greater initial overhead.

The two exponents differ because high-end and low-end machines reach their performance ceilings at different points in the memory hierarchy. On the high-end PC, large SIMD units and plentiful cores dispatch arithmetic so quickly that small datasets remain compute-bound - hence the tiny constant C - but as n grows, the working set spills beyond cache and saturates off-chip bandwidth, so each extra row costs progressively more time, yielding the steeper slope $k \approx 1.33$.

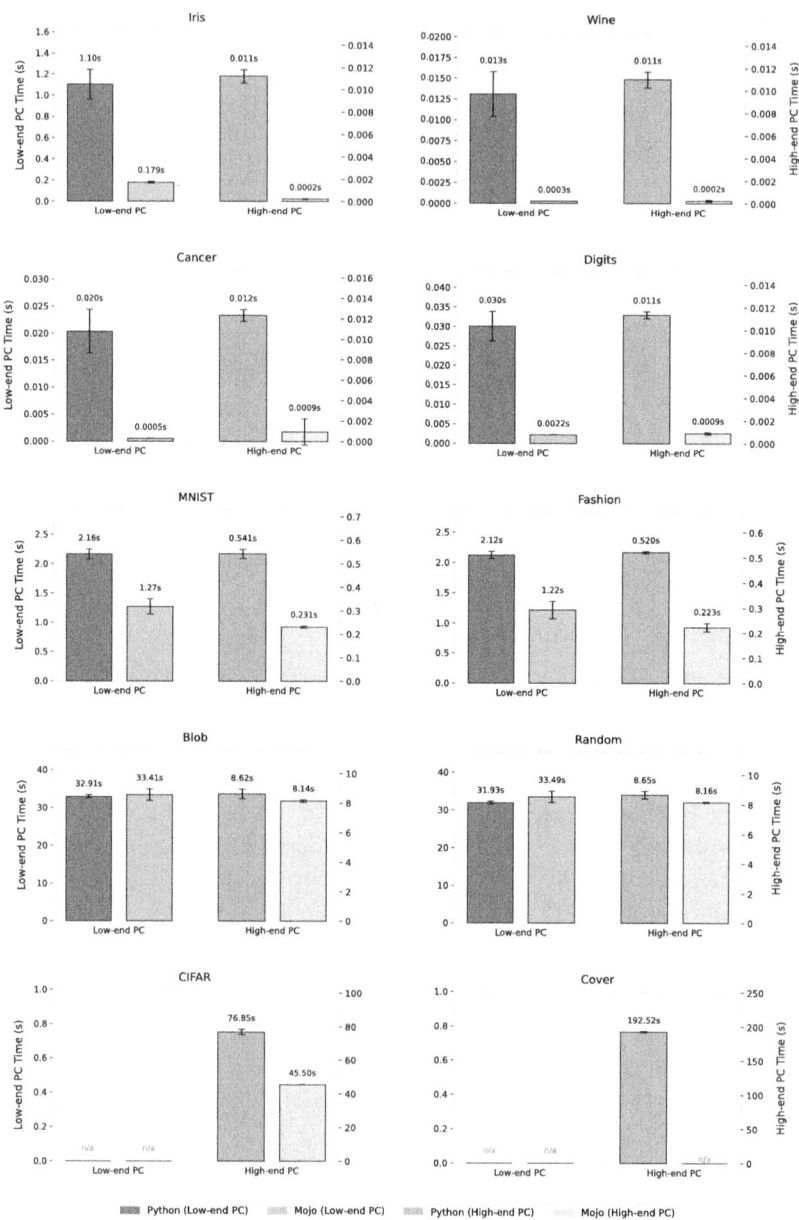

Fig. 3. Performance comparison between Python and Mojo across different datasets and different PCs. Values represent average runtime over ten runs. X-axis is organized by data size, laid out in the methods section.

On the low-end laptop, narrower vectors, fewer cores, and lower bandwidth mean memory stalls dominate almost from the start: the per-row cost rises more slowly in relative terms, so the scaling exponent is smaller ($k \approx 1.19$) even though the baseline overhead C is larger. Thus, the observed k values reflect how quickly each platform changes from compute to memory-bound execution as the dataset grows.

4 Discussion

Our empirical evaluation of Mojo for k-NN acceleration reveals a nuanced performance landscape, offering valuable insights into its practical applicability and current maturity. This section synthesizes these findings, discusses their broader implications, acknowledges the limitations encountered, and outlines avenues for future research.

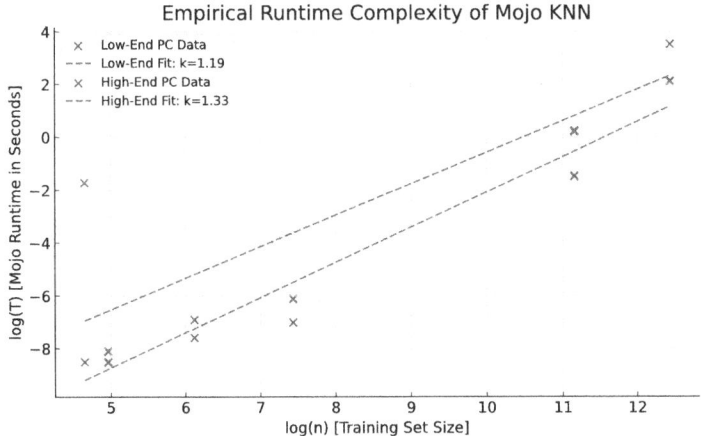

Fig. 4. Empirical runtime complexity of Mojo KNN based on log-log regression. The fitted slopes suggest $O(n^{1.33})$ scaling on high-end and $O(n^{1.19})$ on low-end systems. Constants: $C_{\text{high}} \approx 2.07 \times 10^{-7}$, $C_{\text{low}} \approx 3.80 \times 10^{-6}$.

4.1 Performance Interpretation

Mojo's most significant performance advantages manifest when the k-NN workload is predominantly 'compute bound'. For small to medium-sized tabular datasets (Iris, Wine, Cancer, Digits, MNIST, Fashion-MNIST) and for datasets with very wide feature vectors (CIFAR-10), our compiled SIMD-aware Mojo kernel consistently outperformed scikit-learn by substantial margins, achieving runtime reductions of 40–98%.

In these scenarios, the fixed overhead associated with Python's interpreter, C-API dispatch mechanisms, and scikit-learn's reliance on generic BLAS calls becomes a significant portion of the latter's total execution time. In contrast, Mojo's ahead-of-time compilation and explicit vectorization allow its inner loops to more effectively saturate the CPU's floating-point pipelines.

However, as the number of training samples scales into hundreds of thousands (e.g., the "Blob" and "Random" synthetic datasets), the performance bottleneck often transitions from arithmetic throughput to *memory system efficiency*. On the resource-constrained laptop class CPU, Mojo's computational units appeared to stall more frequently due to cache failures or memory bandwidth limitations, thus eroding its lead and, in some cases, resulting in a slight slowdown compared to scikit-learn.

The workstation's larger last-level caches (LLC) and superior memory bandwidth mitigated these effects, restoring a modest 5–6% performance advantage for Mojo. This highlights that Mojo's speedups are most reliably retained when the active working set of the k-NN algorithm either fits comfortably within the LLC or when the feature vectors are extremely wide (as with CIFAR-10), making each cache line fetch highly productive. Once the memory hierarchy is saturated on constrained hardware, a highly tuned NumPy/BLAS stack, benefitting from decades of optimization, remains a competitive baseline.

4.2 Experimental and Implementation Limitations

Several operational constraints and implementation-specific limitations were observed:

- *Hardware memory constraints:* The low-end laptop's 16 GB RAM proved insufficient for the full CIFAR-10 and Covertype datasets when combined with the working buffers required by either the Mojo or scikit-learn implementations.
- *Mojo kernel scalability for large data:* Our current Mojo k-NN kernel was unable to process the Covertype benchmark on the high-end workstation (approx. 523,000 samples × 54 features). This failure is attributed to the current in-memory design of the kernel, which lacks out-of-core processing capabilities (e.g., memory-mapped I/O or iterative data chunking). This is an implementation-specific bottleneck rather than an inherent limitation of the Mojo language itself. Scikit-learn, using NumPy's mature memory management, successfully handled this dataset.
- *Study scope:* This study focused exclusively on brute-force k-NN on CPUs. The performance characteristics might differ with approximate nearest neighbor algorithms or when leveraging GPU acceleration, which were outside the current scope.

4.3 Practical and Deployment Considerations

The benchmark data suggest three primary "sweet spots" for deploying Mojo-accelerated k-NN:

1. *Latency-critical inference on modest data volumes:* Applications such as real-time recommendation systems or edge analytics, typically processing tens of thousands of records, stand to gain order-of-magnitude speedups with relatively straightforward Mojo implementations.
2. *High-dimensional feature spaces:* Workloads common in computer vision or genomics, which operate on feature vectors with thousands of dimensions (10^3-10^4), can significantly benefit from Mojo's explicit SIMD control and memory layout optimizations.
3. *Compute-rich server environments:* Given sufficient CPU cores and memory bandwidth, Mojo's design for parallelism (e.g., through thread-local contentious buffers and lock-free scheduling) allows for near-linear scaling, making it suitable for server-side, high-throughput inference.

However, for scenarios dominated by extremely large datasets that exceed physical memory, or where the stability and extensive ecosystem of Python/scikit-learn are paramount, the latter remains the more pragmatic choice until Mojo's data handling capabilities and ecosystem mature further.

4.4 Mojo Ecosystem: Early-Stage Caveats and Outlook

Mojo is still pre-1.0, so its APIs change quickly, tooling and documentation are thin, and community support is small. Notebook integration is rudimentary and slows interactive work. Native libraries remain limited, so very large data, fault-tolerant pipelines, or specialized hardware may need extra internal code. Yet, the speed-ups shown here indicate that, as the ecosystem matures, these hurdles should shrink, and Mojo could become a strong option for performance-critical ML.

5 Conclusion

This work delivers the first end-to-end evaluation of a hand-tuned, SIMD vectorized and multithreaded k-NN implementation in Mojo. Across eleven datasets that span five orders of magnitude in size and three in dimensionality, Mojo outran scikit-learn's brute-force classifier by up to 90 % on structured, medium-scale problems and retained a 40 % lead on the high-dimensional CIFAR-10 benchmark. These gains stem from ahead-of-time compilation, explicit memory alignment and direct control over SIMD and thread scheduling—factors that remove Python's interpreter overhead and keep the CPU's floating-point pipelines saturated.

When working sets approach or exceed the last-level cache, however, memory-bandwidth limits narrow the gap, and on a resource-constrained laptop Mojo forfeits its advantage on the largest tables. The study therefore positions Mojo as a compelling choice for latency-sensitive inference on modestly sized or very wide feature spaces, while confirming that mature Python/Cython stacks remain competitive once datasets saturate the memory hierarchy.

Next steps include adding memory-mapped I/O and minibatched processing to scale to 10^7-sample corpora; porting exact and approximate indices (kd-tree, ball-tree, IVF, HNSW) to verify that Mojo's low-level control delivers comparable speed-ups for hierarchical or graph-based search [19,20]; and releasing a self-contained benchmark harness with portable data loaders to spur community replication. We also plan to target heterogeneous back-ends, GPUs and cloud FPGAs, and to extend the study to other classic learners (e.g. SVMs, k mean, gradient boosting) and latency-critical domains such as high frequency trading market discovery and high-dimensional mislabeled learning, testing whether Mojo's performance profile generalizes across the wider spectrum of AI workloads [21–24].

Acknowledgments. The two authors express their gratitude to the Data Science and Artificial Intelligence Innovation Lab at Baylor University, under the leadership of Dr. Henry Han, for providing the resources and support necessary to carry out this exciting research. We also acknowledge the contributions of the Mojo developer community, whose voluntary efforts in addressing questions on GitHub and publishing insightful articles were invaluable during the development process. This work is partially supported by NASA Grant 80NSSC22K1015, NSF 2229138, and McCollum endowed chair startup fund.

Code Availability. The source code for the Mojo KNN implementation, experimental scripts, and benchmark data is publicly available on our GitHub repository: https://github.com/sumanthkolli03/MOJOKNN/tree/master.

References

1. Cover, T., Hart, P.: Nearest neighbor pattern classification. IEEE Trans. Inf. Theory **13**(1), 21–27 (1967). https://doi.org/10.1109/TIT.1967.1053964
2. Uhlmann, J.K.: Satisfying general proximity/similarity queries with metric trees. Inf. Process. Lett. **40**(4), 175–179 (1991). https://doi.org/10.1016/0020-0190(91)90032-N
3. Pedregosa, F., et al.: Scikit-learn: machine learning in python. J. Mach. Learn. Res. **12**, 2825–2830 (2011)
4. Lattner, C.: Introducing Mojo: A Programming Language for All AI Developers. Modular Inc. (2023). https://www.modular.com/blog/mojo-programming-language. Accessed 09 May 2025
5. Sivanandhan, S.: Mojo programming language—68000× faster than python. Medium blog post (2023). https://medium.com/p/d162740a2f67. Accessed 09 May 2025
6. Huang, T., et al.: MojoFrame: a dataframe library in mojo language. arXiv preprint arXiv:2505.04080 (2025)
7. Lichman, M.: UCI machine learning repository. University of California, Irvine, School of Information and Computer Sciences (2013). http://archive.ics.uci.edu/ml
8. LeCun, Y., Bottou, L., Bengio, Y., Haffner, P.: Gradient-based learning applied to document recognition. Proc. IEEE **86**(11), 2278–2324 (1998)

9. Xiao, H., Rasul, K., Vollgraf, R.: Fashion-MNIST: a novel image dataset for benchmarking machine-learning algorithms. arXiv preprint arXiv:1708.07747 (2017)
10. Krizhevsky, A.: Learning multiple layers of features from tiny images. Technical report, University of Toronto (2009)
11. Intel Corporation. Intel® 64 and IA^{-32} Architectures Optimization Reference Manual. Order No. 248966-041 (2023)
12. Mussabayev, R.: Optimizing Euclidean distance computation. Mathematics **12**(23), 1–36 (2024)
13. Wong, S., Duarte, F., Vassiliadis, S.: A Hardware Cache `memcpy` accelerator. In: Proceedings of the 2005 International Conference on Computer Design (ICCD), pp. 234–239. IEEE (2005)
14. Josuttis, N.M.: The C++ Standard Library: A Tutorial and Reference, 2nd edn. Addison–Wesley Professional (2012). §17.6 ("Partial Sorting: `nth_element`")
15. Yu, S., Shah, N., Aggarwal, C.C., Singh, A.K.: ParChain: a framework for parallel hierarchical agglomerative clustering. Proc. VLDB Endow. (PVLDB) **15**(12), 3658–3670 (2022)
16. Harris, C.R., et al.: Array programming with NumPy. Nature **585**(7825), 357–362 (2020)
17. Georges, A., Buytaert, D., Eeckhout, L.: Statistically rigorous Java performance evaluation. In: Proceedings of the 22nd ACM SIGPLAN Conference on Object-Oriented Programming Systems and Applications (OOPSLA), pp. 57–76. ACM (2007). https://doi.org/10.1145/1297027.1297033
18. Johnson, J., Douze, M., Jégou, H.: Billion-scale similarity search with GPUs. IEEE Trans. Big Data **7**(3), 535–547 (2021). https://doi.org/10.1109/TBDATA.2019.2921572
19. Li, H., et al.: Constructing tree-based index for efficient and effective dense retrieval. In: Proceedings of the 46th ACM SIGIR Conference on Research and Development in Information Retrieval (SIGIR 2023), pp. 131–140 (2023). https://doi.org/10.1145/3539618.3591651
20. Aumüller, M., Bernhardsson, E., Faithfull, A.: ANN-benchmarks: a benchmarking tool for approximate nearest neighbor algorithms. Inf. Syst. **87**, 101767 (2020). https://doi.org/10.1016/j.is.2019.02.006
21. Han, H., Li, D., Liu, W., Zhang, H., Wang, J.: High dimensional mislabeled learning. Neurocomputing **573**, 127218 (2024)
22. Han, H., Wu, Y., Wang, J., Han, A.: Interpretable machine learning assessment. Neurocomputing **561**, 126891 (2023)
23. Han, H., Teng, J., Xia, J., Wang, Y., Guo, Z., Li, D.: Predict high-frequency trading marker via manifold learning. Knowl.-Based Syst. **213**, 106662 (2021)
24. Han, H.: Diagnostic biases in translational bioinformatics. BMC Med. Genom. **8**, 1–17 (2015)

A Practical Comparison of Bayesian Computing Platforms in R

Evan Miyakawa[ID] and David Kahle[✉][ID]

Baylor University, Waco, TX 76798, USA
david_kahle@baylor.edu

Abstract. Supported by recent advances in computational tools, Bayesian statistical methods are gaining popularity. This article compares four popular Bayesian computing platforms in R—JAGS, NIMBLE, Stan, and greta—focusing on their methodologies, efficiency, and accuracy. A simulation study evaluates their performance across various models, highlighting JAGS for simpler cases, Stan for complex models, and NIMBLE for flexibility. We provide insights to guide users in selecting the most suitable platform for their needs.

Keywords: Bayesian computation · R · Stan · JAGS · NIMBLE · **greta**

1 Introduction

Bayesian statistical methods are steadily increasing in popularity. Empowered by recent advances in computational algorithms and general purpose real world implementations, more statisticians and data scientists are becoming familiar with both Bayesian statistics and the software used to conduct Bayesian data analysis. Complex hierarchical Bayesian models, for example, can now be fit with only a little more effort than simple models, thanks to the simplicity of Markov chain Monte Carlo (MCMC) methods and their implementations in popular computing platforms.

The abundance of options in Bayesian computing implementations leads to a new dilemma: where to start. How should one choose a Bayesian computing solution? This article presents a systematic review of the Bayesian computing platforms available to the R user, including novel simulations for timing and accuracy considerations.

The article proceeds as follows. In Sect. 2, we give a brief summary of the Bayes paradigm and contextualize the need for Bayesian computing. We also present the basic workings of MCMC methods used in Bayesian inference, highlighting pros and cons of Gibbs sampling and Hamiltonian Monte Carlo (HMC) algorithms, two of the most popular sampling strategies. In Sect. 3, we give an overview of five specific platforms for Bayesian computing: OpenBUGS, JAGS, NIMBLE, Stan, and **greta**. In Sect. 4, we compare the performance of each platform, using a simulation study to look at both computation time and posterior estimation accuracy. Section 5 concludes with a discussion of the results.

2 Monte Carlo Methods for Bayesian Computation

2.1 The Bayes Paradigm in Brief

Bayesian statistical inference is concerned with understanding aspects of a posterior distribution $p(\boldsymbol{\theta}|\mathcal{D})$ over parameters $\boldsymbol{\theta} \in \Theta$ of a statistical model given a prior distribution $p(\boldsymbol{\theta})$ and a likelihood $\ell(\boldsymbol{\theta}|\mathcal{D})$ for those parameters based on our data $\mathcal{D}$. In this article we consider only the continuous scenario; however, the mathematics works for arbitrary measures, and implementations often work in both the continuous and discrete settings. Bayes' theorem states that the posterior distribution of $\boldsymbol{\theta}$ is proportional to the likelihood times the prior:

$$p(\boldsymbol{\theta}|\mathcal{D}) = \frac{\ell(\boldsymbol{\theta}|\mathcal{D})p(\boldsymbol{\theta})}{\int \ell(\boldsymbol{\theta}|\mathcal{D})p(\boldsymbol{\theta})d\boldsymbol{\theta}} \propto \ell(\boldsymbol{\theta}|\mathcal{D})p(\boldsymbol{\theta}), \qquad \boldsymbol{\theta} \in \Theta. \qquad (1)$$

As a classic example, if $X_1, \ldots, X_n \stackrel{iid}{\sim}$ Bernoulli(θ) for a known n, a common prior distribution for $\theta \in \Theta = (0,1)$ would be a Beta(α, β) distribution. Applying Bayes' theorem,

$$\begin{aligned} p(\theta|\mathcal{D}) = p(\theta|x_1, \cdots, x_n) &\propto \theta^{\sum x_i}(1-\theta)^{n-\sum x_i} \frac{\Gamma(\alpha+\beta)}{\Gamma(\alpha)+\Gamma(\beta)}\theta^{\alpha-1}(1-\theta)^{\beta-1} \\ &\propto \theta^{(\sum x_i)+\alpha-1}(1-\theta)^{n-(\sum x_i)+\beta-1}, \end{aligned} \qquad (2)$$

where sums range from $i=1$ to n. Therefore we obtain a closed-form expression of the posterior distribution of θ as a Beta$\left(\alpha + \sum x_i, \beta + \left(n - \sum x_i\right)\right)$. This is, of course, rare: it is very special to be able to determine a closed form for the posterior distribution.

One challenge in Bayesian inference comes in evaluating or otherwise avoiding the integral in the denominator of Bayes' theorem. Without computational tools, if we don't have distributions for the likelihood and prior that are conjugate, closed-form expressions of the posterior are typically not available, and computations respective of the posterior are usually much more challenging.

The need for Bayesian computational methods arises in our search for an alternative approach to understanding the posterior. Most of the time, the goal of Bayesian analysis is not to find the posterior distribution in the form above but to calculate expectations with respect to it, for example its mean, variance, or various probabilities [17]. To accomplish this we typically use Monte Carlo methods: draw large samples from the posterior and approximate the expectations using law of large numbers (LLN) type arguments. The challenge posed by such a strategy is obtaining the draws in the first place.

2.2 Markov Chain Monte Carlo

Markov Chain Monte Carlo (MCMC) methods are commonly used to sample a target distribution, in our case the posterior distribution of a Bayesian analysis. MCMC is a family of algorithms where draws are obtained sequentially via a process akin to a random walk in the sample space of the target distribution.

The process takes the form of a Markov chain, where the distribution of each draw is conditioned only on the preceding value of the chain. Ideally, as the process continues, the marginal distribution of the draws converges to the true target distribution. Therefore, if we run the Markov chain for many iterations, we can expect to see draws coming from the true posterior distribution we are trying to sample. While other strategies exist, they are generally limited relative to MCMC, and as a consequence MCMC has become the go-to algorithm for computing aspects of the posterior distribution [2,9,27].

2.3 Gibbs Sampling

Gibbs sampling (GS) is one of the most common MCMC algorithms used for Bayesian computation. A special case of the Metropolis-Hastings algorithm seen next, the Gibbs sampler works by generating a Markov chain through a series of sequential draws from conditional distributions. Starting from an initial value $\boldsymbol{\theta}^{(0)} = [\theta_1^{(0)}, \ldots, \theta_p^{(0)}]$, the scheme works by sampling $\theta_1^{(1)} \sim p(\theta_1 | \theta_2^{(0)}, \ldots, \theta_p^{(0)}, \mathcal{D})$, $\theta_2^{(1)} \sim p(\theta_2 | \theta_1^{(1)}, \theta_3^{(0)}, \ldots, \theta_p^{(0)}, \mathcal{D})$, and so on up until $\theta_p^{(1)} \sim p(\theta_2 | \theta_1^{(1)}, \theta_2^{(1)}, \ldots, \theta_{p-1}^{(1)}, \mathcal{D})$ to complete $\boldsymbol{\theta}^{(1)}$, and then repeats the process until a given number of draws $\boldsymbol{\theta}^{(1)}, \ldots, \boldsymbol{\theta}^{(N)}$ are determined. Typically blocks of parameters can be sampled at once so that the procedure need not be accomplished one number at a time. When a conditional distribution presents that cannot be sampled in a straightforward way (e.g. as a recognizable distribution or via inverse transform sampling), a more general procedure such as slice sampling or adaptive rejection sampling is typically used.

The Gibbs sampler has a number of advantages. First, it is generally easy to implement. Second, it is efficient with its computations. As we will see with the Metropolis-Hastings and HMC algorithms next, many MCMC samplers reject proposed draws, resulting in seemingly wasted computation. Third, it is surprisingly fast, as we will see in the simulations section.

However, the Gibbs sampler suffers from one major drawback: its intermediate transitions are always along the coordinate axes (unless sampled in blocks, of course). If several variables are highly correlated, traversing the support of the posterior can take a very high amount of iterations because the step-size must be small in order to not step outside high probability regions of the conditional distribution. Since each intermediate transition only allows us to move in one dimension at a time, it may take many steps in order to explore the parameter space.

2.4 Metropolis-Hastings and Hamiltonian Monte Carlo

The question with all MCMC is how to construct a Markov chain with the desired properties: a pre-specified stationary distribution (here the posterior), easy to sample, etc. The Gibbs sampler is one solution to this problem, but it is by no means the only one. Perhaps the most widely applicable MCMC strategy to do so is provided by the Metropolis-Hastings algorithm. The Metropolis-Hastings (MH) algorithm describes a general approach to constructing a Markov chain by specifying a transition kernel from one draw to the next along with a probabilistic correction criterion. The algorithm has two steps: a proposal step, where a new draw is posited, and a correction step, where the draw is accepted or rejected probabilistically based on the relative likelihoods of the current and proposed positions scaled by their relative transition likelihoods [27].

To sample from the posterior $p(\boldsymbol{\theta}|\mathcal{D})$, the MH algorithm requires an initial point $\boldsymbol{\theta}^{(0)} \in \Theta$ and a transition kernel $k(\boldsymbol{\theta}'|\boldsymbol{\theta})$, a distribution over Θ dependent on a given value $\boldsymbol{\theta}$ that is easily sampled from. Starting from $\boldsymbol{\theta}^{(0)}$, a candidate value for $\boldsymbol{\theta}^{(1)}$ is determined by a single draw $\boldsymbol{\theta}'$ from $k(\boldsymbol{\theta}'|\boldsymbol{\theta})$; this is the first step. In the second step, the candidate $\boldsymbol{\theta}'$ is accepted as $\boldsymbol{\theta}^{(1)}$ with probability min $\left(1, \frac{p(\boldsymbol{\theta}'|\mathcal{D})}{p(\boldsymbol{\theta}^{(0)}|\mathcal{D})} \frac{k(\boldsymbol{\theta}^{(0)}|\boldsymbol{\theta}')}{k(\boldsymbol{\theta}'|\boldsymbol{\theta}^{(0)})}\right)$. This process is then repeated for each subsequent draw, typically until some pre-specified number of draws is obtained.

While the scheme is very flexible–it is nearly trivial to select such a transition kernel k–the flexibility also poses a challenge, since the distribution largely influences the performance of the algorithm [13]. Ideally, the distribution should be easy to sample from and allow for easy computation of the acceptance probability. It should also strike a balance between each proposed jump traveling a reasonable distance from the current point while not having proposals rejected too frequently [10]. In practice, the selection of good kernel is the core challenge of a good MH algorithm. Selecting a poor kernel results in draws that are almost always rejected, and unfortunately as the dimension of the parameter space grows the problem of selecting a good proposal distribution becomes even more challenging [1]. This brings us to Hamiltonian Monte Carlo.

Hamiltonian Monte Carlo (HMC) uses the knowledge of the geometry of the target distribution to guide sampling efforts. Under HMC, the density of the target distribution is translated into a potential energy function used to simulate Hamiltonian dynamics to yield Monte Carlo draws. In essence, the proposed value is determined by a physics simulation that mimics a puck sliding around frictionlessly on a surface for a given period of time, stopping it, and using its position as a proposal value. The connection to the posterior distribution is that the surface is defined by the negative log of the posterior distribution, so lower regions of the surface, where the puck will have a tendency to stay in, correspond to regions of higher probability. While the mathematics are technical, the basic concepts and mathematical description are not too challenging; we direct the reader to the excellent introduction [21] in lieu of a longer discussion. It is worth pointing out that the physics simulation involves parameters that themselves need to be set for the algorithm to work well, and this is the role of the No-U-Turn Sampler (NUTS), which can be considered to be a kind of HMC variant [14].

Hamiltonian Monte Carlo algorithms are very robust, in that the Markov chains get stuck in complex regions of the posterior less frequently than other common MCMC methods. When the Markov chains do get stuck, it is easy for the program running the sampler to spot the issues and report them to the user.

Figure 1 gives a visual example of Hamiltonian Monte Carlo sampling versus Gibbs sampling, with 10,000 draws coming from a distribution focused around a complex algebraic pattern defined by Lissajous polynomials [15,16]. Whereas the Gibbs sampler was only able to traverse parts of the posterior distribution and recover only bits and pieces of the pattern, the HMC sampler easily journeyed over the entire posterior distribution and shows the whole pattern.

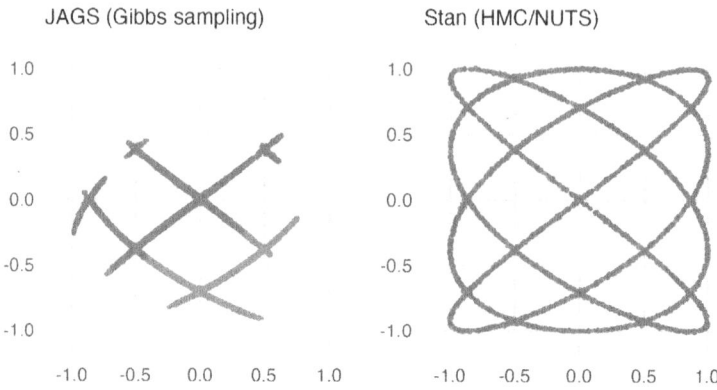

Fig. 1. 10,000 draws (total) sampled using Gibbs sampling and HMC from a distribution centered around a degree-8 polynomial using 8 independent chains. HMC explores the distribution much more rapidly because it is adapted to the geometry of the distribution.

3 Platforms for Bayesian Sampling and Inference

The term probabilistic programming language (PPL) in this article refers to programming languages and associated environments and computational engines that describe probabilistic models and do Bayesian inference more generally. Within Bayesian computing there are several PPLs that integrate with R. This article surveys several of the most common of these platforms used in R with the objective of portraying the strengths, weaknesses, and unique features of each [26]. After describing some parameters common to all or most of these PPLs, in this section we then discuss each platform in detail.

3.1 Common Bayesian Computation Parameters

In this section we describe some common techniques used with MCMC and their related parameters.

Chains. One of the common challenges to MCMC procedures is that their draws tend to be dependent, making inference more challenging as they require more sophisticated LLNs. One strategy to producing more independent draws and better exploring the posterior is to initialize many chains run independently from over-dispersed locations. This strategy also confers a few other advantages: the chains can be easily run in parallel, and the comparison of the chains can be used as a diagnostic device [25]. All the PPLs considered in this work accept a parameter specifying the desired number of chains to run.

Iterations. The number of total draws that each Markov chain collects is a very important setting. If each chain is not set up to run long enough, estimates could be biased due to limited exploration of the posterior or high dependence. Checking convergence statistics can be helpful in diagnosing both of these. An argument specifying the number of iterations for each chain can be inputted in the function call for each sampler.

Warmup. While distinct, the terms "warm-up" and "burn-in" refer to a period of draws at the beginning of the chain that are discarded for being for some reason unsuitable for analysis. Burn-in generally refers to draws not yet believed to have converged to regions of high probability of the posterior, while warm-up is typically used to refer to draws obtained while the sampling procedure is still auto-tuning, e.g. in the NUTS algorithm. Whatever the case, these draws are typically discarded. Each platform accepts an argument like this, though their interpretation varies slightly.

Thinning. In some cases "thinning" is used, in an effort to reduce the dependence among the draws. Thinning refers to discarding some number of draws between those retained, such as keeping every 10th draw. While each sampler defaults to including every simulated posterior draw, a user can specify a "thinning rate", which is the rate at which a sampler will keep memory of the observations. A thinning rate of 1 means that every observation is kept, and a rate of 10 indicates that only every 10th observation will be used.

3.2 OpenBUGS

OpenBUGS is the oldest of the Bayesian sampling platforms that we use, maintained only as recently as 2014 [19]. OpenBUGS is a program that runs exclusively on Windows machines, but there are well-known workarounds allowing users to run it on a Unix-alike machines such as emulators or a virtual machines. OpenBUGS can be used through a point-and-click graphical user interface (GUI) or through R using the **R2OpenBUGS** package, which allows for the user to send model scripts and data to OpenBUGS without needing to leave the R interface [32]. OpenBUGS is based on the BUGS model specification language, which is a small-to-moderately sized set of functions and commands that allow for extensive specification of Bayesian models. The BUGS language is declarative, meaning that order in which expressions are listed in the model code does not matter [18,29].

BUGS stands for Bayesian Analysis Using Gibbs Sampling, and as its name suggests, it uses Gibbs sampling methods by default, but also allows for Metropolis-Hastings and slice sampling. Depending on the model type, OpenBUGS attempts to group nodes that are correlated together and update them at the same time in each Gibbs sampling transition. In most cases this is not feasible, and parameter nodes are updated one at a time, which can be very inefficient if there are many parameters or there is a high level of correlation between them [19,30].

There are a few well known weaknesses to OpenBUGS. The first is the awkward workarounds for non-Windows users. Another is that the program is prone to crashing on occasion, and diagnosing errors in the model code can be cryptic. Unlike a full programming language, OpenBUGS does not control flows such as the if-else statement, although there are workarounds, such as the step() function that is built-in [18]. Also, when using the OpenBUGS application directly, specifying the data in the proper format in order to be read in correctly can be tedious, especially if the data is not very small in size or design structure. For example, if the data was in a .csv file, using R to load a data file and format it into a list is much easier than copying and pasting.

3.3 JAGS

JAGS (Just Another Gibbs Sampler; https://mcmc-jags.sourceforge.io) is similar in purpose to OpenBUGS, in that it enables Bayesian data analysis through posterior sampling using MCMC methods and is based on the BUGS model specification language [22]. However, JAGS is designed to be used interactively entirely in R using the R package **rjags**, making for a more seamless user experience [24]. Other interfaces exist as well. For example, **runjags** is an optional wrapper package that allows for the JAGS model script to be input as a string instead of a separate text file; it also enables running chains in parallel on multiple processes. The **R2jags** wrapper package is also popular, partly because it can automatically stop the MCMC sampling once the chains converge. JAGS itself is implemented in C++, which makes running the MCMC sampling process very fast. Sampling diagnostics can be visualized through the **coda** or **posterior** packages [25,34]. There is plenty of documentation and support available for JAGS, which is still maintained [23]. The documentation includes an installation and user manual, as well as over a dozen examples across dozens of script files.

3.4 NIMBLE

A much newer platform, NIMBLE (https://r-nimble.org) allows for Bayesian computation in R like JAGS and OpenBUGS but is also made for more flexible statistical modeling [6]. It generally uses a BUGS-like specification language, which it cross compiles to C++ under the hood for fasst sampling. Using the **nimble** package, NIMBLE can perform MCMC sampling for Bayesian scenarios using common algorithms, such as Gibbs sampling, but also allows for other types of samplers to be loaded, like slice sampling, reversible jump MCMC, HMC (through **nimbleHMC**), sequential Monte Carlo (through **nimbleSMC**), and even custom-built samplers to be used [7,20,33]. It can also compile other types of algorithms without the BUGS syntax, written in a familiar fashion for R users. The steps to using or creating these algorithms is thoroughly discussed in the NIMBLE user manual, a 250 page document accessible both online and as a PDF, as well as a convenient cheat sheet.

3.5 Stan

Stan (https://mc-stan.org) is another popular PPL for Bayesian computing in R [5]. Like BUGS, Stan has its own unique syntax. Stan has two key differences from the BUGS-based implementations: it has its own language and it uses HMC/NUTS algorithms by default. Stan provides bindings R, Python, and others, including a command line version.

With a look and feel similar to C++, which it cross-compiles into, the Stan language is designed to be more flexible and expressive than the BUGS-type languages. Explicit variable declarations are required in Stan, and unlike the BUGS PPLs, statements are executed in the order that they appear in the model file. It also supports more general purpose programming techniques, such as conditional statements using if-else syntax.

Stan uses HMC by default, which in many cases is more efficient and robust than GS or other MH strategies, which can reduce the computational time for many types of statistical models. Like NIMBLE, Stan model code is translated into C++ and compiled for faster sampling at run time. In addition, Stan conveniently gives the user the option to store a compiled model in a file, which allows for repeated use without the need to re-compile each time with different data, and many IDEs provide syntax highlighting for Stan code.

In R, Stan is used through the **rstan** package [31].[1] The biggest practical difference in using Stan compared to other packages is in the model specification process. Stan requires a model object in the form of a .stan file. This file, either created through a character string in R or in a separate text file, is input along with the data and various control parameters into the stan() function, which performs the sampling. Stan provides a very extensive set of documentation available, including many papers written on the subject, giving Stan users many sources for support. The documentation comes in the form of a user guide, reference manual, and functions reference, each of which is very high quality; these are available online and as downloadable PDFs and are generated by subversion, e.g. 2.35 and 2.36.

3.6 greta

greta (https://greta-stats.org) is a still more recent addition made specifically for easy use in R [11]. The model and data specification process is intended to be as straightforward as possible for R users, who don't need to be familiar with a distinct specification language, unlike other Bayesian platforms: all the coding is done in R itself. Not only does it utilize HMC methods for sampling, but it also uses Google's machine learning platform TensorFlow to perform its

---

[1] Stan can also be used through the newer **cmdstanr** interface; however, we will not describe or analyze that interface in this work [8].

computations. This allows for more flexible use of CPUs and GPUs. As of this writing, **greta** is still in early stages of development and lacks an appropriate amount of documentation and support for users compared to the other implementations mentioned previously. The installation process is far from seamless and is a real barrier to entry as it requires knowledge of Python installations, virtual environments, and the command line. That being said, the project does have a high quality webpage with explicit install instructions, so we expect it to improve over time. While the webpage contains both a function reference and examples, these are only available on the webpage and are much thinner than the massive amounts of documentation of NIMBLE and Stan.

3.7 Other Tools for Specific Scenarios

The Bayesian MCMC platforms mentioned so far cover a lot of the needs in Bayesian analysis. However, there are tools that exist for specific modeling scenarios that can be more useful and readily applicable. Two are particularly noteworthy here: the **rstanarm** and **brms** R packages [3,12]. These are widely used by the R community and built to work very similarly to native R functions such as lm() and glm() but from a Bayesian perspective. Both of these platforms provide a convenient mechanism for specifying priors on classes of parameters, e.g. model coefficients and hierarchical coefficients, although the user does not have as much control as if they were coding in Stan, on which both are based. In the case of **rstanarm**, the models that the package supports are pre-compiled during installation so that the user will never have to wait for model compilation during use.

4 Simulation Study

In this section we conduct several large simulations to assess the relative performance of the PPLs in practical settings. There are several important performance aspects to consider when assessing Bayesian computational strategies. The main two that we focus on are computation time and accuracy. A reliable sampler should both be efficient in its computations and give trustworthy posterior draws. We have conducted simulations to benchmark JAGS, NIMBLE, Stan, and **greta** to compare their performance in each area. It should be noted that we were unable to include OpenBUGS in these simulations, because we wanted to use the exact same machine for each sampling platform, and the machine we used for many of the computations (MacOS Catalina) was unable to use OpenBUGS without emulation.

4.1 Computation Time

In order to assess each sampler's performance across many different scenarios, we chose 16 different model types on which to measure the computation time performance, including simple conjugate models, regression models, hierarchical models, time series models, survival models, and others. For each platform, we ran computations on each model, both for a small dataset and a large dataset (sizes varied by model). Each model computation was done a total of 200 times for each implementation. In all computations, the MCMC sampling was performed with four chains, using 10,000 posterior draws, with a warm-up/burn-in of 1000 iterations prior to sampling, with the exception of **greta** (see below). Sampling was computed in parallel, on the same workstation iMac (2017, 8 GB 2400 MHz DDR4 Memory, 3.4 GHz Quad-Core Intel Core i5 Processor, running MacOS Catalina). With the exceptions of the Stan and **greta** modifications below, to the extent possible we allowed each sampling platform to use its default configurations. Results are contained in Fig. 2.

Stan Particulars. We purposefully decided to benchmark Stan after precompiling and saving each model object, since any Stan model can be saved after compilation and loaded almost instantly for future use. Initial model compilation for Stan took almost exactly 17 s for every model type on our machine.

greta *Particulars.* We also chose to divide **greta**'s sampling across more chains than the other platforms, since **greta** is able to sample on any number of chains, no matter how many cores are available. We used 50 chains for each **greta** computation time simulations, with the same warm-up/burn-in of 1000 iterations, and then 800 samples for each chain, which adds up to 40,000 total draws, the same number as obtained for the other samplers.

We first focus on the computation completion time simulations for 16 models, each using relatively small datasets, to compare platform performance in more basic, every day use cases. Figure 2 graphs the computation time simulations in the form of strip charts. JAGS was consistently the fastest across the board, followed by Stan. NIMBLE and **greta** were notably slower than the first two platforms, which a lot more variation in completion time as well. There were many cases where NIMBLE took 25 times as long to complete sampling compared to JAGS.

We next look at results from a few more complex models, with larger datasets and more intricate parameter spaces to compare platform performance in more heavy use cases. Figure 3 graphs the computation time simulations in the form of strip charts. These results were considerably different than results from more basic models. Stan was usually the fastest in most scenarios and was consistently near the top. **greta** also fared very well in most situations, besides one or two where it was considerably slower. NIMBLE was somewhere in the middle in terms of computation time, but was often in the same range as Stan. JAGS had a noticeable drop-off in computation time performance for many models and was the slowest in most of the scenarios.

One other way of assessing MCMC efficiency is in the form of the effective sample size (ESS) for a posterior distribution. An efficient MCMC algorithm will have a much higher ESS from the same number of MCMC draws, compared to another less efficient algorithm using the same number of draws. One can calculate the ESS for a parameter's posterior, divided by the MCMC computation time, and use that metric as a comparison benchmark between Bayesian MCMC

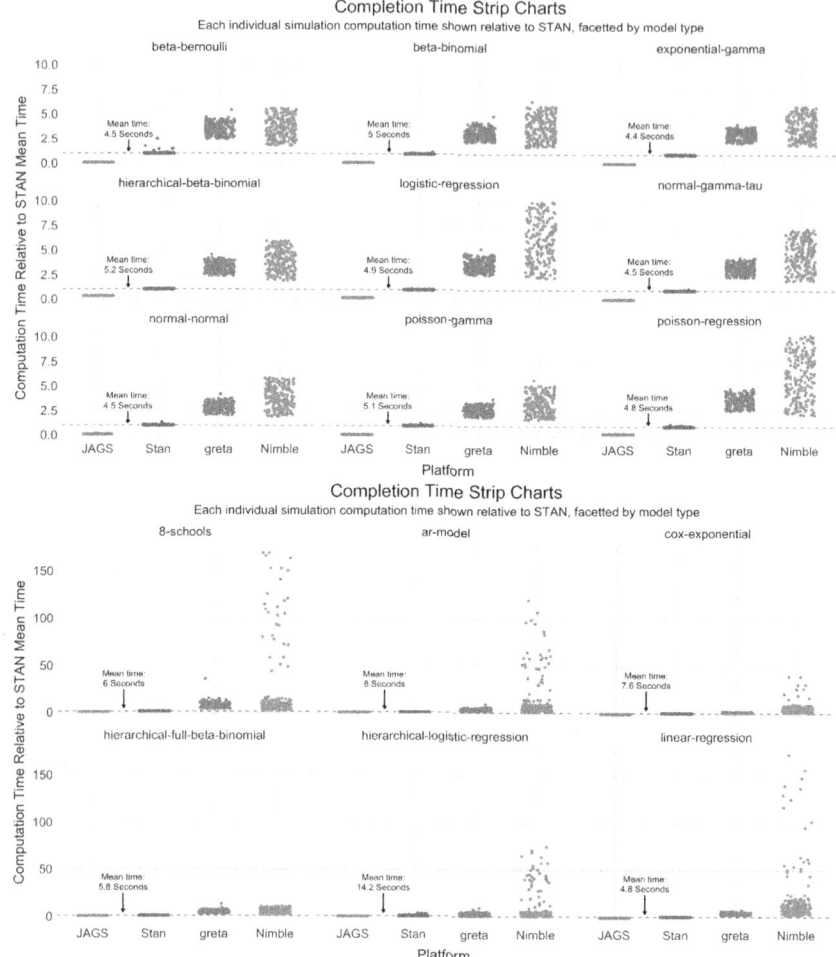

Fig. 2. Individual run times in seconds for each MCMC platform for each of the 16 models with relatively small datasets. Each point represents the completion time for one simulation.

platforms. As an alternative not considered here, designing a simulation study using ESS would be a helpful piece in gaining a better understanding of how the different platforms compare in terms of computational efficiency, and this has been suggested in the Stan community [4].

4.2 Posterior Estimation Accuracy

When using MCMC methods to approximate a posterior distribution, the samples that we collect should ideally be coming from the posterior itself. While we cannot determine if a single draw is from that distribution, there are ways to

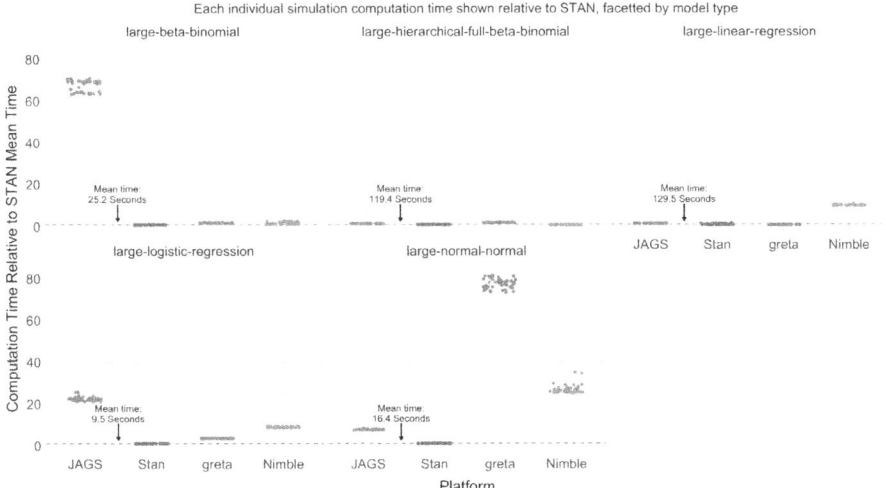

Fig. 3. Run times in seconds for each MCMC platform for six models with larger datasets or parameter spaces. Each point represents the completion time for one simulation.

aggregate all of the draws and see how closely the empirical distribution matches the target posterior distribution. For certain conjugate probability models, we can analytically calculate the posterior distribution for the parameters of interest. In these cases, we use the total variation distance to measure the distance between the true posterior distribution and the estimated posterior distribution, in the form of an empirical distribution obtained from MCMC sampling.

Total variation distance (TVD) is a metric that defines the distance between two probability measures on the same space. For two probability measures P and Q, the TVD is defined $\delta(P,Q) = \sup_{A \in B} |P(A) - Q(A)|$. If P and Q admit densities with respect to Lebesgue measure on $\mathbb{R}$, a well-known result allows for the TVD to be characterized in terms of an integral of the distributions' densities: $\delta(P,Q) = \frac{1}{2} \int_{\mathbb{R}} |p(x) - q(x)| \, dx$.

When our probability models allow us to calculate the analytical posterior distribution, we can use the TVD to calculate the distance between the true distribution and the estimated one. Using this method, we can compare the posterior estimation accuracy of different Bayesian computation implementations.

The **distr** and **distrEx** R packages have a set of functions for working with distributions and computing distances between them, including the `TotalVarDist()` function, which computes the total variation distance of two distributions [28]. This function can handle many different cases, but we will be comparing a known continuous distribution (such as a $\mathcal{N}(0,1)$), and an empirical distribution, specified as a numeric vector of draws from the distribution and estimated inside the function.

To compare the posterior estimation accuracy of our five implementations, we picked six different data models with conjugate prior distributions, where we can analytically derive the true posterior distribution. We then take randomly generated data and calculate the true posterior distribution for that data. After this, for each of OpenBUGS, JAGS, NIMBLE, Stan, and **greta**, we ran a set of 50 simulations using the same data, each time running the MCMC sampler with four chains, 10,000 iterations, and a warm-up or burn-in of 1000, and obtained the samples from the empirical posterior for each to calculate the Total Variation Distance between the true posterior and empirical posterior. For each model, we have aggregated the TVD values for each implementation.

The TVD simulation results are shown in Fig. 4. JAGS, NIMBLE, and Stan all performed very similarly, while **greta** had a lot more variation in TVD, and was generally slightly poorer, since larger Total Variation Distances indicate a posterior estimation that is further from the true posterior.

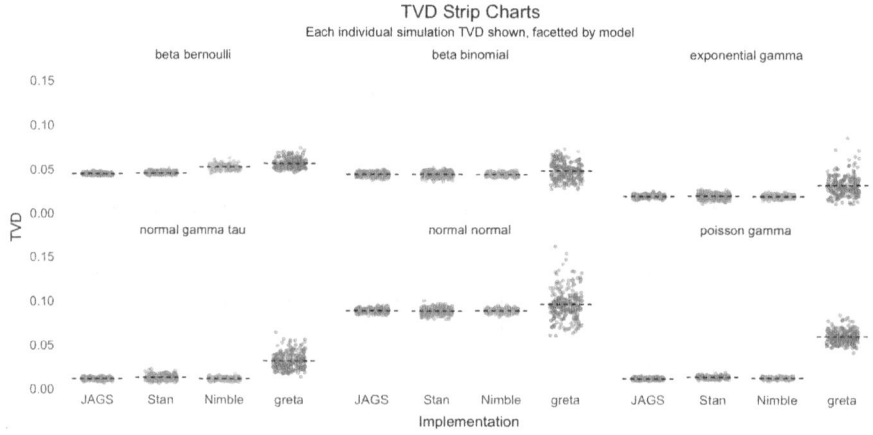

Fig. 4. TVDs for each MCMC platform for six conjugate data models. Each point represents the TVD for one simulation. Dashed lines represent the mean TVD for a given method and model.

Next, we did the same simulations, but using a much smaller number of MCMC iterations per chain (1,000, instead of 10,000). It is even more obvious in 5 that **greta** struggled to have the same posterior estimation accuracy compared to the other platforms.

4.3 Posterior Estimation for Non-conjugate Models

In most cases where we do not have a conjugate data distribution and prior distribution, an analytic posterior cannot be computed. However, most real Bayesian analysis projects require more complex model structures, and measuring the posterior estimation accuracy in these scenarios would also be beneficial. In future

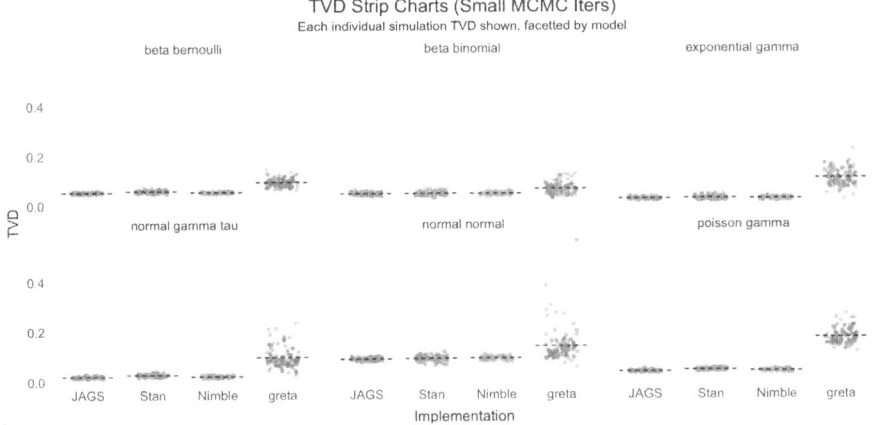

Fig. 5. TVDs for each MCMC platform for six conjugate data models using a smaller number of MCMC iterations.

work, we could obtain an approximation of the true posterior by using Hamiltonian Monte Carlo (through Stan) to get a very large number of draws (way more than needed) from the posterior distribution. We can then treat this empirical posterior distribution as the "true" distribution and compare it to empirical posteriors from OpenBUGS, JAGS, NIMBLE, Stan, **greta**, when they are used with much smaller iterations.

5 Discussion

When it comes down to choosing the "best" Bayesian computing platform, there is no straightforward answer, because each has its own strengths that can make it the best option, depending on the scenario. Here we will highlight what we perceive to be the biggest strengths for each Bayesian computing platform that we have examined, along with the most obvious drawbacks.

JAGS has quickly become a popular choice for Bayesian statisticians, especially those migrating from OpenBUGS, given the almost identical model specification language between the two. Of all the platforms we study in this article, JAGS is the easiest to install and run right out of the box. It is also often the fastest for quick modeling and sampling, and has the least potential hurdles when trying to start doing MCMC sampling quickly. The C++ compilation time when using JAGS is pretty much negligible. Also importantly, despite there being more sophisticated samplers out there, JAGS is still in widespread use. This seems reasonable given JAGS was the fastest in situations where Gibbs sampling could be completed very quickly, given small-to-moderate data sizes and a more straightforward parameter space from which to sample.

The main drawbacks for JAGS come with inherent Gibbs sampling weaknesses. Gibbs sampling is very fast for simpler models and small to medium sized datasets, but it can struggle with more complex parameter spaces or large data sets. It's important to use MCMC sampling diagnostics with all Bayesian computing platforms, but especially when using Gibbs methods.

NIMBLE's main advantage is its flexible algorithm specification and built-in algorithm library. While NIMBLE uses Gibbs sampling methods by default, the possibilities are endless when it comes to other types of computational algorithms that can be chosen or custom-built. NIMBLE also can perform very well relative to other Bayesian computing platforms when dealing with more complex modeling scenarios. Once a NIMBLE model is compiled, the sampling process can be very fast, running on C++ under the hood. NIMBLE also has a very dedicated development team, and it's documentation and community support grows by the year.

While NIMBLE's algorithm capabilities might be appealing to some, the vast majority of Bayesian statisticians may find NIMBLE inferior to JAGS as a straightforward Gibbs Sampler. NIMBLE is slower than JAGS for many common modeling scenarios, considering the amount of time it takes for the model to compile in C++ each time before sampling starts.

Stan seems to be the most robust Bayesian computing platform we experimented with and always seemed to perform well relative to the other sampling tools, regardless of the modeling scenario. Its unique specification language requires more thorough and interpretable coding practices, which makes reading and understanding Stan models very easy. Stan's biggest strength is probably the fact that it uses HMC sampling methods, which is usually way more efficient and reliable. Problem models that could cause issues with Gibbs sampling often were handled with ease in Stan using HMC, and when there were problems, Stan was able to notify the user that there were issues.

While Stan can be a great tool, the initial process of installing it and learning how it works for a new user can be slow. The installation steps for Stan are a bit more complex than other platforms. The Stan model specification language is also very different than BUGS-style programming. Learning how to write Stan models can be an adjustment for Bayesian statisticians already comfortable with the BUGS syntax, although AI now makes translating between the two quite simple. Finally, Stan does take a considerable amount of time to compile a model every time it is edited, which can lead to a slower workflow for someone who is tweaking a model multiple times in a row.

greta is the least well-known of the platforms but has one key advantage: TensorFlow based parallelization. **greta** is also has the most intuitive model specification for R users who don't already know BUGS or Stan syntax. **greta**-style programming looks almost identical to usual R programming, which offers a quick on-ramp to Bayesian computing for R users who haven't done it before.

greta has a few drawbacks that may keep it from growing quickly in popularity, at least at present. Installing **greta** can be a laborious process. **greta** relies on Python being already built, but it isn't uncommon for the underlying Python installation to cause issues when configuring **greta** to work in R. At the time of our testing **greta** also had the least robust documentation of all of the platforms, which means finding answers to common problems is more difficult. Due to it's unique model specification, some specific model types can be much harder to program.

References

1. Betancourt, M.: A conceptual introduction to Hamiltonian Monte Carlo. arXiv preprint arXiv:1701.02434 (2017)
2. Brooks, S., Gelman, A., Jones, G., Meng, X.L. (eds.): Handbook of Markov Chain Monte Carlo. Chapman and Hall/CRC, Boca Raton (2011)
3. Bürkner, P.C.: brms: an R package for Bayesian multilevel models using Stan. J. Stat. Softw. **80**(1), 1–28 (2017). https://doi.org/10.18637/jss.v080.i01
4. Carpenter, B.: Re: how to compare different algorithms for Bayesian inference in terms of speed (and subject to effectiveness) (2018). https://tinyurl.com/3hfzshju
5. Carpenter, B., et al.: Stan: a probabilistic programming language. J. Stat. Softw. **76**(1), 1–32 (2017). https://doi.org/10.18637/jss.v076.i01
6. de Valpine, P., Turek, D., Paciorek, C., Anderson-Bergman, C., Temple Lang, D., Bodik, R.: Programming with models: writing statistical algorithms for general model structures with NIMBLE. J. Comput. Graph. Stat. **26**, 403–417 (2017). https://doi.org/10.1080/10618600.2016.1172487
7. de Valpine, P., et al.: NIMBLE user manual (2024). https://doi.org/10.5281/zenodo.1211190, https://r-nimble.org. R package manual version 1.3.0
8. Gabry, J., Češnovar, R., Johnson, A., Bronder, S.: cmdstanr: R interface to 'CmdStan' (2024). https://mc-stan.org/cmdstanr/, https://discourse.mc-stan.org. R package version 0.8.1
9. Gamerman, D., Lopes, H.F.: Markov Chain Monte Carlo: Stochastic Simulation for Bayesian Inference. CRC Press (2006)
10. Gelman, A., Carlin, J.B., Stern, H.S., Dunson, D.B., Vehtari, A., Rubin, D.B.: Bayesian Data Analysis, 3 edn. Chapman and Hall/CRC, Boca Raton (2013). https://doi.org/10.1201/b16018
11. Golding, N.: greta: simple and scalable statistical modelling in R. J. Open Sour. Softw. **4**(40), 1601 (2019). https://doi.org/10.21105/joss.01601
12. Goodrich, B., Gabry, J., Ali, I., Brilleman, S.: rstanarm: Bayesian applied regression modeling via Stan. (2024). https://mc-stan.org/rstanarm/. R package version 2.32.1
13. Green, P.J., Łatuszyński, K., Pereyra, M., Robert, C.P.: Bayesian computation: a summary of the current state, and samples backwards and forwards. Stat. Comput. **25**(4), 835–862 (2015). https://doi.org/10.1007/s11222-015-9574-5
14. Hoffman, M.D., Gelman, A.: The No-U-turn sampler: adaptively setting path lengths in Hamiltonian Monte Carlo. J. Mach. Learn. Res. **15**(47), 1593–1623 (2014). https://www.jmlr.org/papers/volume15/hoffman14a/hoffman14a.pdf
15. Kahle, D.: mpoly: multivariate polynomials in R. R J. **5**(1), 162–170 (2013). https://journal.r-project.org/archive/2013-1/kahle.pdf
16. Kahle, D., Hauenstein, J.D.: Stochastic exploration of real varieties via variety distributions (2024). https://arxiv.org/abs/2410.16071
17. Kahle, D.J., Seaman, J.W., Stamey, J.D.: An overview of Bayesian computation. In: Faya, P., Pourmohamad, T. (eds.) Case Studies in Bayesian Methods for Biopharmaceutical CMC, pp. 5–24. Chapman & Hall/CRC Biostatistics, Chapman and Hall/CRC (2022)

18. Lunn, D., Spiegelhalter, D., Thomas, A., Best, N.: The BUGS project: evolution, critique and future directions. Stat. Med. **28**(25), 3049–3067 (2009). https://doi.org/10.1002/sim.3680
19. Lunn, D.J., Thomas, A., Best, N., Spiegelhalter, D.: WinBUGS - a Bayesian modelling framework: concepts, structure, and extensibility. Stat. Comput. **10**(4), 325–337 (2000). https://doi.org/10.1023/A:1008929526011
20. Michaud, N., de Valpine, P., Turek, D., Paciorek, C.J., Nguyen, D.: Sequential Monte Carlo methods in the nimble and nimbleSMC R packages. J. Stat. Softw. **100**, 1–39 (2021)
21. Neal, R.M.: MCMC using Hamiltonian dynamics. In: Brooks, S., Gelman, A., Jones, G., Meng, X.L. (eds.) Handbook of Markov Chain Monte Carlo, chap. 11, pp. 113–162. Chapman and Hall/CRC (2011)
22. Plummer, M.: JAGS: a program for analysis of Bayesian graphical models using Gibbs sampling. In: Proceedings of the 3rd International Workshop on Distributed Statistical Computing (DSC 2003), Vienna, Austria (2003). https://www.r-project.org/conferences/DSC-2003/Proceedings/Plummer.pdf
23. Plummer, M.: JAGS Version 4.3.0 user manual (2017)
24. Plummer, M.: rjags: Bayesian Graphical Models using MCMC (2024). https://CRAN.R-project.org/package=rjags. R package version 4-16
25. Plummer, M., Best, N., Cowles, K., Vines, K.: CODA: convergence diagnosis and output analysis for MCMC. R News **6**(1), 7–11 (2006). https://www.r-project.org/doc/Rnews/Rnews_2006-1.pdf
26. R Core Team: R: A Language and Environment for Statistical Computing. R Foundation for Statistical Computing, Vienna, Austria (2024). https://www.R-project.org/
27. Robert, C.P., Casella, G.: Monte Carlo Statistical Methods, 2nd edn. Springer, New York (2004)
28. Ruckdeschel, P., Kohl, M., Stabla, T., Camphausen, F.: S4 classes for distributions. R News **6**(2), 2–6 (2006)
29. Spiegelhalter, D., Thomas, A., Best, N., Lunn, D.: WinBUGS User Manual, version 1.4 edn. (2003)
30. Spiegelhalter, D., Thomas, A., Best, N., Lunn, D.: OpenBUGS User Manual, version 3.2.3 edn. (2014)
31. Stan Development Team: RStan: the R interface to Stan (2020). http://mc-stan.org/. R package version 2.21.2
32. Sturtz, S., Ligges, U., Gelman, A.: R2WinBUGS: a package for running WinBUGS from R. J. Stat. Softw. **12**(3), 1–16 (2005) https://doi.org/10.18637/jss.v012.i03
33. Turek, D., de Valpine, P., Paciorek, C.: nimbleHMC: Hamiltonian Monte Carlo and other gradient-based MCMC sampling algorithms for 'nimble' (2024). R package version 0.2.3
34. Vehtari, A., Gelman, A., Simpson, D., Carpenter, B., Bürkner, P.C.: Rank-normalization, folding, and localization: an improved Rhat for assessing convergence of MCMC (with discussion). Bayesian Anal. **16**(2), 667–718 (2021)

Attention-Enhanced Deep Learning

CATC-Net: A CoAttention-Guided Temporal Capsule Network for Speech Emotion Recognition

Yuanyuan Wei[1(✉)] and Heming Huang[2(✉)]

[1] School of Computer, Qinghai Normal University, No. 38 Wusi West Road, Chengxi District, Xining 810008, China
[2] The State Key Laboratory of Tibetan Intelligence, Xining 810008, China
huanghm@qhnu.edu.cn

Abstract. Speech emotion recognition is a crucial technology in human-computer interaction. However, it faces such challenges as insufficient feature selection, limited temporal modeling capability, and high complexity of emotional expression. To address these issues, a novel model called CoAttention-guided Temporal Capsule Network (CATC-Net) is proposed. It integrates a collaborative attention mechanism, bidirectional scalable long short-term memory network (Bi-sLSTM), and capsule networks for comprehensive modeling of multi-dimensional emotional features. The model begins with a feature desensitization mapping block to enhance generalization in complex and noisy environments. Next, a channel-wise weighting strategy based on multi-head attention is introduced to jointly model and fuse global contextual information with channel-level importance. The Bi-sLSTM structure further improves the capture of speech rhythm and dynamic emotional changes. Finally, the capsule network compresses and encodes high-level semantic features while preserving their compositional relationships. Extensive experiments on datasets IEMOCAP, EMODB, CASIA, and BodEMODB show that the proposed CATC-Net achieves higher accuracy than existing methods, demonstrating strong robustness and effective emotion recognition.

Keywords: Speech Emotion Recognition · Collaborative Attention · Bi-sLSTM

1 Introduction

Speech, as a key medium for human communication, conveys rich emotional information and plays an important role in human-computer interaction (HCI) [1]. Speech emotion recognition (SER) aims to identify emotions conveyed by speakers through vocal signals. Today, this technology has found wide application in various intelligent systems, including driver assistance, psychological state assessment, emotion-aware service robots, and etc.

Traditional SER methods primarily rely on machine learning algorithms to classify emotions from speech signals. Common techniques include support vector machines (SVM) [2], hidden Markov models (HMM) [3], and other related approaches [4]. These methods generally depend on manually extracted features such as Mel-frequency cepstral coefficients (MFCCs), zero-crossing rate, and pitch, and perform emotion recognition based on statistical modeling [5]. However, their limited adaptability to variations in speech signals results in certain constraints on overall performance.

Convolutional neural networks (CNN) can capture local time-frequency patterns from speech feature maps effectively [6]. But CNNs struggle with modeling long-term temporal dependencies. Recurrent neural networks (RNN) are better at capturing temporal information in speech signals and modeling dynamic emotional features over time [7]. However, RNNs face gradient vanishing or explosion issues when training long sequences. Long short-term memory networks (LSTM) solve gradient issues by using gating mechanisms to retain long-term dependencies, though at the cost of increased complexity and training effort [8]. Gated recurrent units (GRU), a simplified LSTM, have a streamlined structure and faster training. Yet for tasks needing to model complex long-term dependencies, GRUs may underperform compared to LSTM [9].

Recently, attention mechanisms have enhanced SER by enabling models to focus on emotionally relevant speech segments through weighted emphasis [10]. The multi-head mechanism enables parallel learning of diverse features, boosting the model's capacity to capture complex emotional cues [11]. Capsule networks (CapsNet) have been introduced into SER to better preserve emotional feature relationships through capsule-based hierarchical encoding [12, 13].

Despite great advances in SER, key limitations persist. Transformer-based and multi-head attention models mainly focus on temporal dependencies, overlooking the importance of feature dimensions in emotion recognition. Traditional LSTM variants, though capable of modeling sequences, struggle with highly dynamic emotional fluctuations, especially during rapid changes or fast speech rhythms. Moreover, many existing models originate from image or NLP tasks and lack optimization for the unique structure and emotional patterns of speech signals.

Therefore, a CoAttention-guided Temporal Capsule Network, abbreviated as CATC-Net, is proposed for SER, aiming to comprehensively model emotional features. First, a collaborative attention mechanism (CAM) incorporates a channel-wise weighting modulation strategy based on multi-head attention (MHA). Then, a bidirectional scalable LSTM (Bi-sLSTM) enhances temporal modeling by capturing both forward and backward dependencies, improving sensitivity to dynamic cues like rhythm and intonation. Finally, a capsule network (CapsNet) compresses and encodes features while preserving compositional relationships among emotional cues, enhancing robustness to variations such as speech rate and accent. Our main contributions are as follows:

- An innovative CAM module, combined with a Bi-sLSTM structure, is proposed to achieve more comprehensive temporal modeling and feature enhancement for speech emotion features.
- A novel model CATC-Net, integrating a feature desensitization module, CAM, Bi-sLSTM, and CapsNet, is introduced to model features from multiple perspectives, including temporal, feature, and spatial dimensions.
- Extensive experiments on datasets with different languages and scenarios, including IEMOCAP, EMODB, CASIA, and BodEMODB, show that the proposed model CATC-Net achieves outstanding performance across various evaluation metrics.

2 Related Works

As a key HCI technology, SER is challenging due to emotional diversity and variability. Modeling temporal dynamics, feature correlations, and spatial relationships is crucial for better performance.

Early SER methods rely on manually extracted features like MFCC, pitch, and energy, classified by algorithms such as SVM, HMM, or GMM [14]. Bhavan et al. [15] extract spectral features from EMODB, RAVDESS, and IITKGP-SEHSC datasets and use an ensemble of Gaussian kernel SVMs for emotion classification. Nwe et al. [16] extract short-time logarithmic frequency power coefficients (LFPC) and use a Hidden Markov Model (HMM) to classify six emotions. Přibil et al. [17] studies factors affecting GMM-based emotion classification, including parameter initialization and feature ordering. Traditional methods based on manual features and specific classifiers suffer from limited generalization, performance, and stability, especially in complex or multi-class tasks.

Deep learning enables automatic feature extraction and complex pattern recognition, driving major advancements and becoming a key focus in SER research. Trigeorgis et al. [18] combine CNN and LSTM using the database RECOLA to learn speech representations and context-aware emotional features. CNNs capture local patterns well but struggle with global temporal context due to their limited receptive fields. RNNs like LSTM and GRU, with memory cells, better model temporal dependencies and improve the performance of SER.

In SER, attention mechanisms help models focus on key emotional cues in complex sequences, enhancing recognition accuracy where traditional models fall short. Fan et al. [19] propose a hierarchical CNN with posterior attention that captures spatiotemporal information and long-term dependencies, effectively extracting discriminative emotional features for SER. Udeh et al. [20] improve SER using ShuffleNet V2 by converting speech to spectrograms, integrating ECA-Net attention, changing activation functions, and adjusting learning rates. Nfissi et al. [21] proposed an end-to-end multi-resolution SER framework using fast discrete wavelet transform, combining 1D expanded CNN with spatial attention and Bi-GRU with temporal attention to extract emotional features.

CapsNet [22] emerge as a promising deep learning architecture, offering new possibilities for SER tasks due to their ability to accurately model spatial hierarchies and complex patterns. Wu et al. [23] introduce a CapsNet-based SER model that captures spatial relationships in spectrograms and uses recurrent connections, improving emotion recognition accuracy. Building on this, Zhang et al. [12] propose the CENN model combining MHA, residual, and capsule modules to enhance global and local feature extraction, improving SER performance.

3 Methods

CATC-Net is an innovative architecture for SER designed to significantly enhance emotion recognition performance by integrating multiple advanced modules, as shown in Fig. 1. It mainly comprises four key components: a feature desensitization mapping block

(F-DMB), a collaborative attention mechanism with feature recalibration (CAM), a Bi-sLSTM structure with temporal awareness, and a capsule network module (CapsNet) to model high-level semantic relationships.

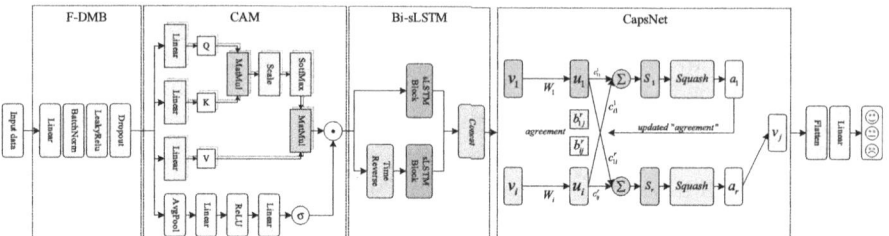

Fig. 1. Overview of the CATC-Net architecture. The model consists of four main modules: the F-DMB module performs preliminary processing on the input speech features to enhance robustness in complex environments; the CAM module extracts key channel and temporal features through a collaborative attention mechanism; the Bi-sLSTM module captures bidirectional temporal dependencies; and the CapsNet module performs high-dimensional modeling and emotion classification using dynamic routing.

For the input multidimensional speech features $X \in \mathbb{R}^{B \times T \times d}$, the model first applies preliminary processing through the F-DMB to adapt to the subsequent network structure and effectively improve generalization in complex environments. Then, the features pass sequentially through the CAM and the Bi-sLSTM structure to recalibrate salient spatial features and effectively model temporal dependencies in the speech sequence. And finally, the extracted and integrated features are fed into the CapsNet module for precise classification of speech emotions.

3.1 Feature Desensitization Mapping

To enhance the generalization and modeling capability of input features, a model called F-DMB is the first model designed to apply linear transformation, normalization, and redistribution. It maps the input features into a high-dimensional space through a linear fully connected layer, breaking explicit correlations among the original features. Simultaneously, BatchNormalization (BN) normalizes the distribution of each channel, alleviating internal covariate shift. Then, the LeakyReLU activation function introduces nonlinear modeling capability, combined with Dropout to prevent overfitting. The overall process is as follow:

$$Y_1 = \text{Dropout}(\phi_{LR}(\text{BN}(\mathbf{W}X + \mathbf{b}))). \tag{1}$$

where, $X \in \mathbb{R}^{B \times T \times d}$ represents the input features and $Y_1 \in \mathbb{R}^{B \times T \times D}$ represents the output features. $\mathbf{W}$ and $\mathbf{b}$ represent the weights and biases of the linear layer, BN(·) denotes batch normalization, and $\phi_{LR}(\cdot)$ stands for the LeakyReLU activation function.

3.2 Collaborative Attention Mechanism

To capture both global dependencies in the feature sequence and important information in the channel dimension, a CAM module is designed. It incorporates a channel-wise weighting modulation strategy based on multi-head attention (MHA), enabling joint modeling and fusion of global contextual information with the importance of channel-level features. Specifically, given the input feature Y_1, the model computes MHA and channel weights separately. The calculation process of MHA proceeds as follows:

$$\mathbf{Q} = \mathbf{W}_q Y_1, \quad \mathbf{K} = \mathbf{W}_k Y_1, \quad \mathbf{V} = \mathbf{W}_v Y_1, \tag{2}$$

$$Y_{mha} = \text{Attention}(\mathbf{Q}, \mathbf{K}, \mathbf{V}) = \text{Softmax}\left(\frac{\mathbf{Q}\mathbf{K}^\top}{\sqrt{d_k}}\right)\mathbf{V}. \tag{3}$$

At the same time, the model computes weights along the channel dimension. It first applies global average pooling to Y_1 to obtain a channel descriptor vector, which is then passed through a bottleneck structure composed of two fully connected layers to generate the channel weights W. The specific calculation process is as follows:

$$W = \sigma(\mathbf{W}_2 \cdot \phi_{Re}(\mathbf{W}_1 \cdot \text{GAP}(Y_1) + \mathbf{b}_1) + \mathbf{b}_2). \tag{4}$$

where, $\text{GAP}(\cdot)$ denotes global average pooling on the feature maps, $\phi_{Re}(\cdot)$ represents the ReLU activation function, $\sigma(\cdot)$ stands for the Sigmoid activation function, $\mathbf{W}_1$ and $\mathbf{W}_2$ are the weights of the linear layers, and $\mathbf{b}_1$ and $\mathbf{b}_2$ are the biases of the linear layers. Finally, the outputs of the two branches are fused through element-wise multiplication:

$$Y_{cam} = W \odot Y_{mha}. \tag{5}$$

where, $\odot$ denotes element-wise multiplication along the channel dimension, enabling joint optimization in both spatial and channel dimensions.

3.3 Bi-sLSTM

Traditional LSTM models, although possessing strong sequence modeling capability, face challenges such as low computational efficiency and limited capacity when dealing with long sequences. To improve the ability of the proposed model to capture multi-scale information within long sequences, sLSTM is adopted as the fundamental modeling unit. It retains the gating mechanisms of the standard LSTM while introducing key components such as exponential gating, normalizer state, and stable state, which effectively enable control over modeling at different temporal scales [48].

However, sLSTM is limited to utilizing only the contextual information preceding the current time step within the sequence, which hinders its ability to fully capture the temporal dependency structure. Therefore, the sLSTM is extended into a bidirectional structure (Bi-sLSTM), which models the input sequence in both forward and backward directions to capture dependencies on both sides of the current time step. In the implementation, the input sequence Y_{cam} is first separately fed into the forward and backward sLSTM units, resulting in:

$$H_{fwd} = \overrightarrow{\text{sLSTM}}(Y_{cam}), \tag{6}$$
$$H_{bwd} = \overleftarrow{\text{sLSTM}}(\text{flip}(Y_{cam})).$$

where, flip(·) denotes sequence reversal. Then, the two outputs are concatenated along the feature dimension to form a representation containing complete contextual information.

$$Y_{bi} = [H_{fwd}; H_{bwd}] \in \mathbb{R}^{B \times T \times 2D}. \tag{7}$$

This module not only retains the efficient modeling capability of sLSTM but also significantly enhances the ability of the model to comprehensively capture contextual information within the time series.

3.4 Capsule Network Module

CapsNet uses dynamic routing to assign features to candidate output capsules, enabling more selective and refined representations. It improves the adaptability of the model to feature distributions and suppresses irrelevant information, enhancing its sensitivity to emotional cues. Thus, CapsNet strengthens the structured semantic representation and improves the modeling of complex speech.

In a capsule network, each lower-level capsule i establishes a connection with an upper-level capsule j through a set of learnable affine transformation matrices $\mathbf{W}_{ij}$, generating a prediction vector:

$$\hat{\mathbf{u}}_{j|i} = \mathbf{W}_{ij}\mathbf{u}_i. \tag{8}$$

where, $\mathbf{u}_i$ is the output vector of the i-th lower-level capsule and $\hat{\mathbf{u}}_{j|i}$ is its prediction for the upper-level capsule j.

The input to the upper-level capsule j is finally computed as the weighted sum of all prediction vectors:

$$\mathbf{s}_j = \sum_i c_{ij}\hat{\mathbf{u}}_{j|i}. \tag{9}$$

where, c_{ij} is the routing coefficient obtained through iterative updates by the dynamic routing mechanism, representing the degree of confidence with which capsule i sends information to capsule j. This value satisfies the normalization constraint $\sum_j c_{ij} = 1$. The output of the upper-level capsule is obtained through a nonlinear squashing function:

$$\mathbf{v}_j = \frac{\mathbf{s}_j^2}{1+\mathbf{s}_j^2} \cdot \frac{\mathbf{s}_j}{\mathbf{s}_j}. \tag{10}$$

This function ensures that the length of the capsule output vector lies within the range (0, 1), representing the probability of the existence of a specific pattern.

4 Experimental

To comprehensively evaluate the proposed CATC-Net, experiments are conducted on four SER datasets: IEMOCAP [24], CASIA [25], EMODB [26], and BodEMODB. The dataset IEMOCAP contains 1103 anger, 1084 sadness, 1636 happiness, and 1708 neutral samples. CASIA and BodEMODB provide 1200 and 3000 samples respectively, evenly distributed across six emotion categories. EMODB includes 535 samples covering anger (127), boredom (81), fear (69), happiness (71), neutral (79), sadness (62), and disgust (46).

The input features used in this study are consistent with the feature set adopted by Zhang et al. [27], representing multidimensional fused audio feature representations. Each speech frame is represented by 110 acoustic features, forming a feature tensor of shape 130 × 110, where 130 is the unified number of time frames. The 110 dimensions include 20 MFCCs, 1 zero crossing rate, 12 chroma features, 7 spectral contrast features, 20 first-order delta MFCCs, 10 spectral centroid features, 10 spectral rolloff features, 20 s-order delta MFCCs, 6 Tonnetz features, 1 spectral bandwidth feature, and 3 empty feature dimensions. By fusing these multi-level acoustic features, the model effectively captures differences in emotion, timbre, and intonation in speech. After feature extraction, all features undergo normalization, which converts each feature dimension to a standard distribution with a mean of 0 and a standard deviation of 1 [28].

4.1 Implementation Details

The experiments are conducted using PyTorch 2.3.0, Python 3.12, and CUDA 12.1 on a system equipped with an NVIDIA GeForce RTX 4090D GPU with 24GB memory. The model is trained with a batch size of 64. The input feature dimension is 130 by 110, which is first mapped to 128 dimensions via a linear layer, followed by BatchNorm1d normalization and a LeakyReLU activation with a negative slope of 0.01. Dropout with a probability of 0.7 is applied to mitigate overfitting. The attention mechanism uses two heads, and the capsule network employs dynamic routing with three iterations. The AdamW optimizer is used with an initial learning rate of 0.001, combined with a CosineAnnealingLR scheduler for smooth learning rate decay. Cross-entropy loss with label smoothing using a smoothing factor of 0.1 is applied to improve generalization. A fixed random seed of 42 is set for reproducibility. The dataset is randomly split with 90% for training and 10% for testing, using 10-fold cross-validation. Each fold is trained for 500 epochs.

4.2 Ablation Study

Table 1 presents the results of an ablation study conducted on the dataset EMODB to analyze the individual contributions of key modules in the CATC-Net model.

A baseline model composed of MHA, sLSTM, and MLP is adopted as a reference, achieving an average classification accuracy of 86.85 ± 3.08%, which serves as a clear benchmark for subsequent comparisons.

Building on the baseline, CAM, Bi-sLSTM, and CapsNet are individually introduced, leading to improved accuracies of 87.96 ± 2.79%, 87.22 ± 2.82%, and 89.07

Table 1. Ablation analysis of the proposed CATC-Net is conducted on the EMODB dataset.

CAM	Bi-sLSTM	CapsNet	Acc ± Std (%)	p-value
			86.85 ± 3.08	0.00066
✓			87.96 ± 2.79	0.00040
	✓		87.22 ± 2.82	0.00026
		✓	89.07 ± 1.84	0.00105
✓		✓	92.41 ± 2.22	0.67930
	✓	✓	90.93 ± 1.37	0.00377
✓	✓		88.35 ± 2.47	0.00166
✓	✓	✓	92.77 ± 1.05	–

± 1.84%, respectively. These results confirm that each module contributes positively to performance through spatial recalibration, temporal modeling, and feature encapsulation. When combining two modules, the CAM and CapsNet combination achieves the highest accuracy of 92.41 ± 2.22%, showing a strong synergistic effect. The combinations of Bi-sLSTM with CapsNet and CAM with Bi-sLSTM also yield improved accuracies of 90.93 ± 1.37% and 88.35% ± 2.47%. Integrating all three modules forms the complete CATC-Net, which reaches the best performance of 92.77 ± 1.05%, with high accuracy and strong stability. Paired t-tests between each variant and the complete model show that removing any single module causes a statistically significant drop in performance, with all corresponding p-values below 0.01. The only exception is the CAM and CapsNet combination, where the performance difference is not statistically significant, suggesting that this pair retains most of the effectiveness of the complete model.

In summary, the ablation study confirms the essential roles and synergy of CAM, Bi-sLSTM, and CapsNet, demonstrating the robustness and superior performance of CATC-Net in SER.

4.3 Evaluation Results

Figure 2 presents the recognition performance of CATC-Net across different emotion categories, and a confusion matrix serves as the evaluation tool.

CATC-Net performs well on clear and high-quality samples across multiple datasets, achieving 100% accuracy for disgust (D) and sadness (Sa) on EMODB, over 90% accuracy for happiness (H), neutral (N), and anger (A) on BodEMODB, and strong results for anger (A), neutral (N), fear (F), and sadness (Sa) on CASIA. However, it struggles with ambiguous or similar emotions, showing high misclassification rates for boredom (B), fear (F), and neutral (N) on EMODB, lower accuracy for fear (F) on BodEMODB, and confusion between surprise (Su) and happiness (H) on CASIA. On the more complex dataset IEMOCAP, overall accuracy is lower, especially for happiness (H), neutral (N), and sadness (Sa), indicating that further improvement is needed in fine-grained emotion discrimination and robustness in complex scenarios.

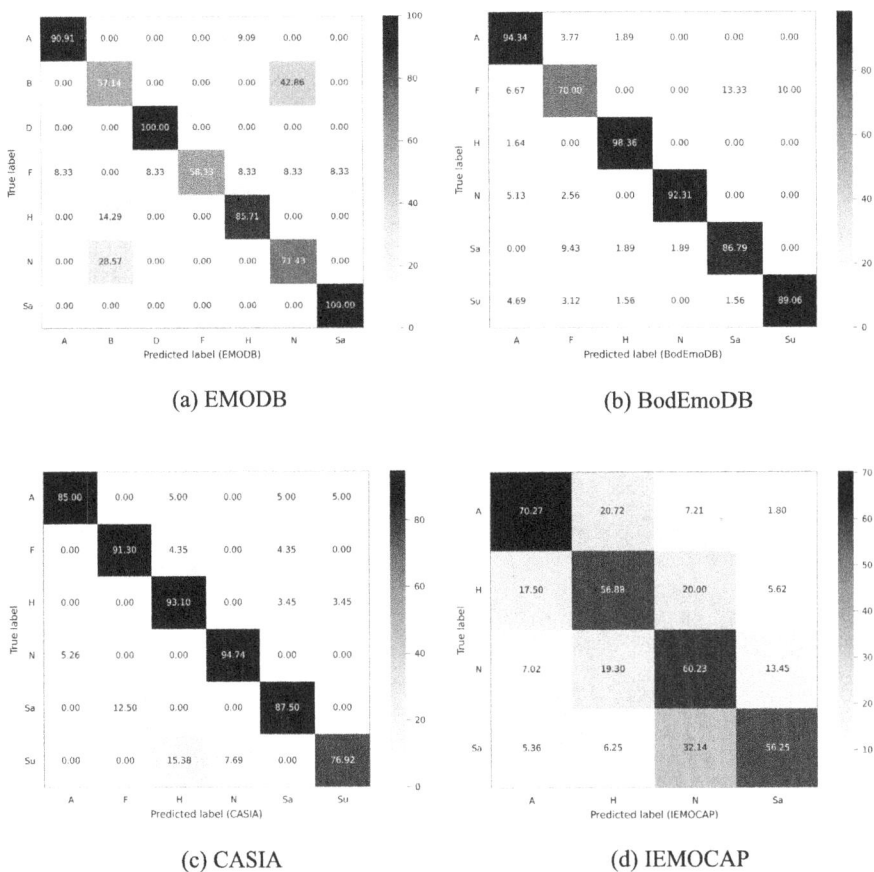

Fig. 2. Confusion matrices for the proposed CATC-Net evaluated on four datasets.

4.4 Performance Comparison with Recent Methods

To validate the effectiveness of the proposed CATC-Net model, performance comparisons with various state-of-the-art (SOTA) models are conducted on four emotional speech datasets: EMODB, IEMOCAP, BodEMODB, and CASIA, as shown in Table 2.

On the dataset EMODB, CATC-Net achieves an accuracy of 92.77%, significantly outperforms mainstream methods such as CENN (90.00%), TIM-Net (89.19%), and DSTCNet (88.79%). The CAM and Bi-sLSTM structure integrated into CATC-Net effectively enhance the capability of the model to capture key emotional cues from high-dimensional speech features. Moreover, by leveraging both forward and backward contextual information, these components improve temporal dependency modeling, allowing the model to maintain strong discriminative power even in scenarios with limited data.

On the dataset IEMOCAP, CATC-Net achieves an accuracy of 67.96%, outperforming methods like MCIL, WADNA + DNN, and Dual-TBNet. This dataset contains natural dialogues with complex speech and background noise. The F-DMB reduces

Table 2. Accuracy comparison between the proposed CATC-Net and SOTA models on the four datasets.

Dataset	Model	Accuracy(%)	Dataset	Model	Accuracy(%)
EMODB	WADAN + DNN [29]	84.49	IEMOCAP	Dual-TBNet [32]	64.80
	MSCRNN-A [30]	88.41		SFF-NEC [34]	64.52
	DSTCNet [31]	88.79		DRP [36]	57.58
	AMSNet [10]	88.34		SpeechFormer ++ [37]	62.90
	Dual-TBNet [32]	84.10		WADNA + DNN [29]	66.92
	TIM-Net [33]	89.19		MCIL [38]	67.00
	SFF-NEC [34]	82.84		Speech Swin-Trans [39]	62.90
	Bi-MGAN [35]	80.16		CubicKD [40]	63.32 ± 1.00
	CENN [12]	90.00		PulseEmoNet [41]	61.40
	CATC-Net	**92.77 ± 1.05**		**CATC-Net**	**67.96 ± 1.00**
Dataset	**Model**	**Accuracy(%)**	**Dataset**	**Model**	**Accuracy(%)**
BodEMODB	LSTM [42]	82.33	CASIA	SVM + DT [45]	85.30
	GRU [43]	77.17		MSCRNN-A [30]	60.75
	CNN [28]	74.17		CASC-AN [46]	67.08
	TCN [44]	78.17		Bi-MGAN [35]	80.41
	CENN [12]	85.00		MVAN-DJL [47]	70.74
	STACN [27]	91.70		MA-CapsNet [13]	76.28
	PulseEmoNet [41]	88.70		hc-former [19]	87.08
	CATC-Net	**91.90 ± 0.65**		**CATC-Net**	**92.75 ± 0.79**

non-emotional interference such as speaker variability and noise, while the Bi-sLSTM models long-term dependencies, allowing the model to effectively capture emotional dynamics.

On the Tibetan emotional speech dataset BodEMODB, CATC-Net achieves an accu-racy of 91.90%, outperforming several classic models such as STACN (91.70%), PulseEmoNet (88.70%), and LSTM (82.33%). This dataset presents significant linguistic differences and cultural diversity, which pose challenges to the transferability and generalization capabilities of the model. CapsNet incorporated in CATC-Net effectively models the "part-whole" relationships among semantic features, allowing the model to accurately capture emotional patterns across different language structures.

On the dataset CASIA, CATC-Net reaches the highest accuracy of 92.75%, outperforming methods like hc-former (87.08%), MA-CapsNet (76.28%), and MVAN-DJL (70.74%). CASIA, with good recordings and standardized expressions, is ideal for training and validating models. In CATC-Net, the F-DMB and CAM modules filter redundant info early and focus on emotion-relevant areas. CapsNet's high-level modeling also boosts the recognition of standardized expressions, resulting in better overall performance.

The proposed CATC-Net performs outstandingly across four emotional speech datasets, outperforming SOTA models. It shows strong generalization and adaptability to various environments. By integrating modules like CAM, Bi-sLSTM, F-DMB, and CapsNet, it can extract emotional features, enhance temporal modeling, and understand emotional constructs, thus achieving stable and efficient recognition even under tough conditions.

4.5 Complexity Analysis

Theoretical Analysis. This section provides a theoretical analysis of the time complexity of CATC-Net. Suppose the input feature tensor has the shape $[B, T, D]$, where B denotes the batch size, T the sequence length, and D the input dimension. The time complexity of each module can be derived as follows:

F-DMB module: The main computational cost arises from linear transformations, resulting in a time complexity of $O(BTDd)$, where d denotes the projected feature dimension. CAM module: This module is primarily composed of a multi-head self-attention mechanism, with a time complexity of $O(BT^2d + BTd^2)$. The first term corresponds to the computation of attention weights, and the second term corresponds to the linear transformation of the value vectors. Bi-sLSTM module: The main computational overhead stems from the bidirectional sLSTM and the feedforward network. The overall time complexity can be summarized as $O(BTd^2)$. CapsNet module: After the Bi-sLSTM module, the input to CapsNet becomes $[B, T, 2d]$. CapsNet performs dynamic routing through low-rank projection, vector norm calculation, and weighted aggregation. Its time complexity is $O(BTN_c d_c r)$, where N_c is the number of capsules, d_c the dimension of each capsule, and r the number of routing iterations. In summary, the total time complexity of the model is given by: $O(BTDd + BT^2d + BTd^2 + BTN_c d_c r)$.

Empirical Analysis. As shown in Table 3, CATC-Net is significantly outperformed in terms of parameter count, computational complexity, and GPU memory usage compared to existing models such as SpeechFormer [49] and SpeechFormer++ [37]. Only

Table 3. Performance and Resource Comparison of CATC-Net and Existing Models.

Metric	CATC-Net	SpeechFormer	SpeechFormer + +
Parameters	**341,975**	17,770,884	17,774,468
Trainable Parameter	**341,975**	16,746,884	16,750,468
FLOPs (G)	**0.0297**	2.2811	2.3972
Inference Latency (ms/sample)	70.51	**7.50**	11.61
GPU Memory Usage (MB)	**26.51**	92.02	100.67

about one-fiftieth of the parameters and less than one-seventieth of the FLOPs are used, demonstrating strong advantages in lightweight design, especially for deployment in resource-constrained environments. The introduction of the Bi-sLSTM module results in an increased inference latency; however, it remains within an acceptable range, and the overall resource efficiency still demonstrates strong competitiveness.

5 Conclusion

In this study, CATC-Net, a CoAttention-guided Temporal Capsule Network, is proposed. It is designed to integrate a channel-wise weighting mechanism based on multi-head attention, a bidirectional scalable long short-term memory network (Bi-sLSTM), and capsule networks (CapsNet), enabling comprehensive modeling of multidimensional emotional features in speech. Extensive experiments show that CATC-Net consistently outperforms SOTA models in both accuracy and robustness.

Building upon related research, the design of CATC-Net further advances the modeling depth and expressive power of SER. Previous studies have combined CNN and LSTM to jointly model local and temporal features of speech. Trigeorgis et al. [18], for example, demonstrate the effectiveness of this approach through their work. As attention mechanisms become increasingly integrated into emotion recognition, Fan et al. [19] propose a hierarchical CNN combined with posterior attention fusion. At the feature representation level, the CapsNet architecture is gaining attention for its advantage in modeling spatial hierarchical relationships. Wu et al. [23] further introduce it into SER tasks, improving the capability of the model to capture complex emotional configurations. CATC-Net advances these approaches by enhancing coordination between spatial and temporal features through integrated attention and capsule routing mechanisms.

Despite the advantages of CATC-Net in recognition accuracy and innovation, it has limitations. It only uses the acoustic modality as input, facing challenges in complex or cross-modal scenarios. Future work could incorporate multimodal info like semantic text and visual signals to improve generalization and context awareness. Also, its parameter size and inference efficiency need optimization. A lightweight version for edge deployment is being explored. So, future efforts can aim at a more compact and faster model while keeping recognition performance for real-time emotional interaction.

Declarations. This research was supported by the Natural Science Foundation of Qinghai Province of China (No. 2022-ZJ-925) and the National Natural Science Foundation of China (No. 62066039).

References

1. O'Brien, H.L., Roll, I., Kampen, A., Davoudi, N.: Rethinking (Dis)engagement in human-computer interaction. Comput. Hum. Behav. **128**, 107109 (2022). https://doi.org/10.1016/j.chb.2021.107109
2. He, X., Lin, L., Deng, J., Wang, L.: Speech emotion recognition based on SVM with local temporal-spectral features. IEEE Access **9**, 112897–112907 (2021)
3. Liu, C., Jia, X., Wu, S., Du, J.: Speech emotion recognition based on HMM and spiking neural network. IEEE Trans. Neural Netw. Learn. Syst. **31**, 1665–1677 (2020)
4. Reddy, B.N., Rani, T.S., Kumar, B.M.: Speech emotion recognition using gaussian mixture model with deep learning techniques. Int. J. Innov. Technol. Explor. Eng. **10**, 2734–2743 (2021)
5. Chen, X., Chen, H., Hu, Y., et al.: Centroid-oriented extracting transform and its application in seismic spectral decomposition. IEEE Trans. Geosci. Remote Sens. **62**, 1–11 (2024)
6. Flower, T.M.L., Jaya, T.: A novel concatenated 1D-CNN model for speech emotion recognition. Biomed. Signal Process. Control **93**, 106201 (2024)
7. Mirsamadi, S., Barsoum, E., Zhang, C.: Automatic speech emotion recognition using recurrent neural networks with local attention. In: 2017 IEEE International Conference on Acoustics, Speech and Signal Processing (ICASSP), pp. 2227–2231. IEEE, New Orleans, LA, USA (2017)
8. Yang, Z., Li, Z., Zhou, S., et al.: Speech emotion recognition based on multi-feature speed rate and LSTM. Neurocomputing **601**, 128177 (2024)
9. Ahmed, M.R., Islam, S., Islam, A.K.M.M., et al.: An ensemble 1D-CNN-LSTM-GRU model with data augmentation for speech emotion recognition. Expert Syst. Appl. **218**, 119633 (2023)
10. Chen, Z., Li, J., Liu, H., et al.: Learning multi-scale features for speech emotion recognition with connection attention mechanism. Expert Syst. Appl. **214**, 118943 (2023)
11. Chen, Z., Lin, M., Wang, Z., et al.: Spatio-temporal representation learning enhanced speech emotion recognition with multi-head attention mechanisms. Knowl.-Based Syst. **281**, 111077 (2023)
12. Zhang, H., Huang, H., Zhao, P., et al.: CENN: Capsule-enhanced neural network with innovative metrics for robust speech emotion recognition. Knowl.-Based Syst. **304**, 112499 (2024)
13. Zhang, H., Huang, H., Han, H.: MA-CapsNet-DA: speech emotion recognition based on MA-CapsNet using data augmentation. Expert Syst. Appl. **244**, 122939 (2024)
14. Patel, M.P., Chaudhari, A.A., Pund, M.A., et al.: Speech emotion recognition system using Gaussian mixture model and improvement proposed via boosted GMM. Development **56**, 64 (2017)
15. Bhavan, A., Chauhan, P., Shah, R.R.: Bagged support vector machines for emotion recognition from speech. Knowl.-Based Syst. **184**, 104886 (2019)
16. Nwe, T.L., Foo, S.W., De Silva, L.C.: Speech emotion recognition using hidden Markov models. Speech Commun. **41**(4), 603–623 (2003)
17. Přibil, J., Přibilová, A.: Evaluation of influence of spectral and prosodic features on GMM classification of Czech and Slovak emotional speech. EURASIP J. Audio Speech Music Process. **2013**, 1–22 (2013)

18. Trigeorgis, G., Ringeval, F., Brueckner, R., et al.: Adieu features? End-to-end speech emotion recognition using a deep convolutional recurrent network. In: IEEE Int. Conf. on Acoustics, Speech and Signal Processing (ICASSP), pp. 5200–5204. IEEE, New Orleans (2016)
19. Fan, Y., Huang, H., Han, H.: Hierarchical convolutional neural networks with post-attention for speech emotion recognition. Neurocomputing **615**, 128879 (2025)
20. Udeh, C.P., Chen, L., Du, S., et al.: Improved ShuffleNet V2 network with attention for speech emotion recognition. Inf. Sci. **689**, 121488 (2025)
21. Nfissi, A., Bouachir, W., Bouguila, N., et al.: SigWavNet: learning multiresolution signal wavelet network for speech emotion recognition. IEEE Trans. Affect. Comput. (2025)
22. Sabour, S., Frosst, N., Hinton, G.E.: Dynamic routing between capsules. Adv. Neural Inf. Process. Syst. **30** (2017)
23. Wu, X., Liu, S., Cao, Y., et al.: Speech emotion recognition using capsule networks. In: IEEE Int. Conf. on Acoustics, Speech and Signal Processing (ICASSP), pp. 6695–6699. IEEE, Brighton (2019)
24. Busso, C., Bulut, M., Lee, C.C., et al.: IEMOCAP: interactive emotional dyadic motion capture database. Lang. Resour. Eval. **42**, 335–359 (2008)
25. Zhang, J., Jia, H.: Design of speech corpus for mandarin text to speech. In: The Blizzard Challenge 2008 Workshop. ISCA, Brisbane (2008)
26. Burkhardt, F., Paeschke, A., Rolfes, M., et al.: A Database of German Emotional Speech. In: INTERSPEECH 2005, pp. 1517–1520. ISCA, Lisbon (2005)
27. Zhang, H., Huang, H., Zhao, P., et al.: Sparse temporal aware capsule network for robust speech emotion recognition. Eng. Appl. Artif. Intell. **144**, 110060 (2025)
28. Peng, Z., Lu, Y., Pan, S., et al.: Efficient speech emotion recognition using multi-scale CNN and attention. In: IEEE Int. Conf. on Acoustics, Speech and Signal Processing (ICASSP), pp. 3020–3024. IEEE, Toronto (2021)
29. Yi, L., Mak, M.W.: Improving speech emotion recognition with adversarial data augmentation network. IEEE Trans. Neural Netw. Learn. Syst. **33**(1), 172–184 (2020)
30. Tao, H., Geng, L., Shan, S., et al.: Multi-stream convolution-recurrent neural networks based on attention mechanism fusion for speech emotion recognition. Entropy **24**(8), 1025 (2022)
31. Guo, L., Ding, S., Wang, L., et al.: Dstcnet: Deep spectro-temporal-channel attention network for speech emotion recognition. IEEE Trans. Neural Netw. Learn. Syst. (2023)
32. Liu, Z., Kang, X., Ren, F.: Dual-TBNet: Improving the robustness of speech features via dual-transformer-BiLSTM for speech emotion recognition. IEEE/ACM Trans. Audio Speech Lang. Process. **31**, 2193–2203 (2023)
33. Ye, J., Wen, X.C., Wei, Y., et al.: Temporal modeling matters: a novel temporal emotional modeling approach for speech emotion recognition. In: IEEE Int. Conf. on Acoustics, Speech and Signal Processing (ICASSP), pp. 1–5. IEEE, Rhodes Island (2023)
34. Thirumuru, R., Gurugubelli, K., Vuppala, A.K.: Novel feature representation using single frequency filtering and nonlinear energy operator for speech emotion recognition. Digit. Signal Process. **120**, 103293 (2022)
35. Li, H., Zhang, X., Duan, S., et al.: Speech emotion recognition based on bi-directional acoustic–articulatory conversion. Knowl. -Based Syst. **299**, 112123 (2024)
36. Guo, L., Wang, L., Dang, J., et al.: Learning affective representations based on magnitude and dynamic relative phase information for speech emotion recognition. Speech Commun. **136**, 118–127 (2022)
37. Chen, W., Xing, X., Xu, X., et al.: Speechformer++: a hierarchical efficient framework for paralinguistic speech processing. IEEE/ACM Trans. Audio Speech Lang. Process. **31**, 775–788 (2023)
38. Zhou, Y., Liang, X., Gu, Y., et al.: Multi-classifier interactive learning for ambiguous speech emotion recognition. IEEE/ACM Trans. Audio Speech Lang. Process. **30**, 695–705 (2022)

39. Wang, Y., Lu, C., Lian, H., et al.: Speech swin-transformer: exploring a hierarchical transformer with shifted windows for speech emotion recognition. In: IEEE Int. Conf. on Acoustics, Speech and Signal Processing (ICASSP), pp. 11646–11650. IEEE, Seoul (2024)
40. Lou, Z., Otake, S., Li, Z., et al.: Cubic knowledge distillation for speech emotion recognition. In: IEEE Int. Conf. on Acoustics, Speech and Signal Processing (ICASSP), pp. 5705–5709. IEEE, Seoul (2024)
41. Zhang, H., Tang, G., Huang, H., et al.: PulseEmoNet: pulse emotion network for speech emotion recognition. Biomed. Signal Process. Control **105**, 107687 (2025)
42. Wang, J., Xue, M., Culhane, R., et al.: Speech emotion recognition with dual-sequence LSTM architecture. In: IEEE Int. Conf. on Acoustics, Speech and Signal Processing (ICASSP), pp. 6474–6478. IEEE, Barcelona (2020)
43. Rajamani, S.T., Rajamani, K.T., Mallol-Ragolta, A., et al.: A novel attention-based gated recurrent unit and its efficacy in speech emotion recognition. In: IEEE Int. Conf. on Acoustics, Speech and Signal Processing (ICASSP), pp. 6294–6298. IEEE, Toronto (2021)
44. Zhang, X., Tang, J., Cao, H., et al.: Cascaded speech separation denoising and dereverberation using attention and TCN-WPE networks for speech devices. IEEE Internet Things J. **11**(10), 18047–18058 (2024)
45. Sun, L., Li, Q., Fu, S., et al.: Speech emotion recognition based on genetic algorithm–decision tree fusion of deep and acoustic features. ETRI J. **44**(3), 462–475 (2022)
46. Liu, Y., Sun, H., Guan, W., et al.: A discriminative feature representation method based on cascaded attention network with adversarial strategy for speech emotion recognition. IEEE/ACM Trans. Audio Speech Lang. Process. **31**, 1063–1074 (2023)
47. Liu, Y., Chen, X., Song, Y., et al.: Discriminative feature learning based on multi-view attention network with diffusion joint loss for speech emotion recognition. Eng. Appl. Artif. Intell. **137**, 109219 (2024)
48. Beck, M., Pöppel, K., Spanring, M., et al.: xlstm: Extended Long Short-Term Memory. arXiv preprint arXiv:2405.04517 (2024)
49. Chen, W., Liu, H., Zhou, Y., et al.: SpeechFormer: A Hierarchical Efficient Framework Incorporating the Characteristics of Speech. arXiv preprint arXiv:2203.03812 (2022)

YOLOv5-TS: Channel-Spatial Attention Guided Bidirectional Feature Fusion for Traffic Sign Detection

Yanbang Zhang[1,2](✉), Fen Zhang[1,2], Yuanfan Gao[1], and Enyan Guo[1]

[1] College of Mathematics and Information Science, Xianyang Normal University, Xianyang 712000, Shaanxi, China
zhangyanbang@163.com
[2] School of Mathematics and Statistics, Xianyang Normal University, Xianyang 712000, Shaanxi, China

Abstract. With the rapid development of autonomous driving technology, the demand for real-time and accurate traffic sign detection has been increasing. To address the issue of missed detections of small objects and occluded signs in complex scenarios, this paper proposes an improved YOLOv5 model based on the fusion of CBAM (Convolutional Block Attention Module) and BiFPN (Bidirectional Feature Pyramid Network). By introducing CBAM to enhance the model's ability to focus on key features and combining it with BiFPN to optimize the efficiency of multi-scale feature fusion, the detection robustness in complex environments is significantly improved. Experiments on the CTSD dataset show that the improved model achieves an average precision (mAP) of 92.5%, which is a 2.1% improvement over the original YOLOv5, while maintaining real-time detection speed (119 FPS). This research provides an efficient solution for the engineering application of traffic sign detection.

Keywords: YOLOv5 · Traffic sign detection · Attention Mechanism · Feature pyramid network

1 Introduction

With the rapid evolution of autonomous driving technology from SAE Level 2 (partial automation) to Level 4 (high automation), the environmental perception systems demand unprecedented precision and real-time performance in traffic sign detection [1]. According to McKinsey's projection, the global autonomous vehicle market is expected to exceed $600 billion by 2030. As standardized visual regulators of road environments, traffic signs provide essential navigation constraints including speed limits, directional instructions, and hazard warnings. Misdetection or misclassification of these signs poses significant safety risks, evidenced by NHTSA's 2022 Autonomous Vehicle Incident Report indicating that 12% of automated driving system-related crashes stemmed from traffic sign recognition errors.

Traffic sign detection technology is mainly divided into traditional methods based on machine learning and modern methods based on deep learning [2]. Traditional methods rely on manually designed features such as color and shape (such as HSV color space segmentation, HOG descriptors) [3–7], but lack robustness in complex lighting, occlusion, and scale changing scenes. In deep learning methods, two-stage algorithms such as Faster R-CNN [8] achieve high-precision detection through region proposal mechanisms, but the computational cost is relatively high. Although single-stage algorithms such as YOLO series [9, 10] and SSD [11] simplify the process and improve speed, they face the challenge of high missed detection rates for small objects.

The development of the YOLO series reflected the evolution of object detection technology [12–14]. YOLOv1 first proposed an end-to-end detection paradigm, which directly predicts bounding boxes and categories through regression, laying the foundation for real-time detection [9]. YOLOv2 introduced batch normalization and Darknet-19 backbone network, utilizing anchor box clustering to optimize prior box design [15]. YOLOv3 enhanced small object detection capability by combining Darknet-53 network with multi-scale prediction [16]. Subsequent versions continued to be optimized, such as YOLOv4 integrating Mosaic data augmentation and cross stage local network (CSP-Darknet53) [17], and YOLOv5 achieving efficient deployment of edge devices through lightweight design [18].

In response to the challenges of complex scenarios, recent research has focused on multi-scale feature optimization and model robustness improvement. Dewi et al. combined Generative Adversarial Networks (GANs) to enhance the diversity of training data [19], while Liu et al. improved the YOLOv3 backbone network to balance speed and accuracy [20]. The Wang team designed the DM-SPCSPC module to enhance multi-scale feature fusion [21], Hou et al. proposed the FL SLKNet network to solve the problem of feature degradation in rainy and foggy environments [22], and Chen et al. optimized long-distance dependency and small object detection through a dual interaction architecture [23]. In the direction of lightweighting, Wei et al. used depthwise separable convolution to reduce computational complexity [24], while Cui et al. implemented efficient multi-scale fusion based on Transformer architecture [5].

The improvement of feature pyramid networks is still a research hotspot. BiFPN improves multi-scale detection performance through bidirectional cross scale connections [17], but existing methods often directly use generic structures and do not design adaptive strategies for the scale characteristics of traffic signs. Although lightweight improvements (such as Ghost module [17]) reduce parameters, they can easily lead to the loss of small target information.

In response to the above issues, this paper proposes an improved YOLOv5 model that integrates CBAM and BiFPN. By embedding CBAM after each C3 module in the CSPDarknet53 backbone network, the synergistic effect of channel spatial attention can be achieved. Experiments have shown that it can suppress background interference and enhance the response of landmark regions. Simultaneously reconstructing the BiFPN structure, introducing cross layer skip connections and learnable weight normalization to solve the problem of imbalanced contribution of multi-scale features, and increasing the precision of small object detection. This paper comprehensively validates the performance of the model on CTSD datasets.

2 Method Design

This section elaborates on the design of an improved YOLOv5 model based on CBAM-BiFPN fusion, encompassing module-level optimization strategies and innovative points in model construction.

2.1 The Structure of the Proposed Model

The YOLOv5-CBAM-BiFPN model proposed in this paper is based on YOLOv5s. By embedding the CBAM in the backbone network and replacing the original PANet with a BiFPN, a detection framework with enhanced multi-scale feature fusion is constructed (Fig. 1).

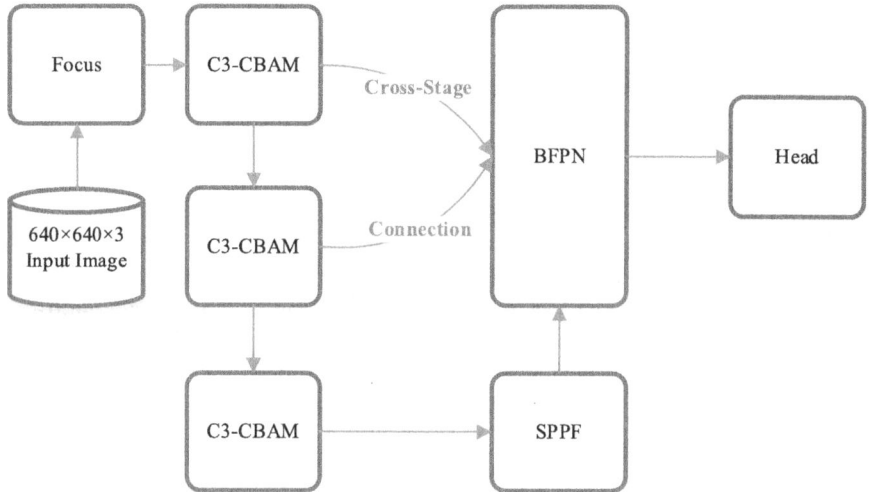

Fig. 1. Overall Framework of the Network Structure

The proposed YOLOv5-CBAM-BiFPN architecture retains the overall framework of YOLOv5s while introducing significant modifications to enhance feature representation and fusion. The backbone network employs a CSPDarknet53 structure, where each C3 module is augmented with a CBAM to dynamically recalibrate channel-wise and spatial features through dual attention mechanisms. This integration effectively suppresses background noise and highlights discriminative regions, improving small object detection performance. The neck network replaces the original PANet with an optimized BiFPN, which incorporates cross-layer skip connections and learnable weight normalization to facilitate bidirectional information flow across scales. By dynamically balancing feature contributions through weighted fusion, the improved BiFPN alleviates semantic gap issues in traditional pyramids. The detection head preserves YOLOv5's three-scale anchor-based prediction mechanism, leveraging multi-level features from the enhanced neck to output class probabilities, bounding box coordinates, and confidence scores. This design maintains real-time inference efficiency while addressing the challenges of scale variation and complex backgrounds in traffic sign detection.

2.2 CBAM Embedding Strategy

2.2.1 Channel-Spatial Attention Collaborative Mechanism

CBAM is composed of Channel Attention Module (CAM) and Spatial Attention Module (SAM), and its core idea is to enhance the response of key regions through dynamic feature calibration. In this algorithm, the CBAM module is embedded behind each C3 module of the YOLOv5 backbone network.

(1) Channel Attention Calculation

Perform global average pooling and maximum pooling on the input feature map $F \in \mathbb{R}^{C \times H \times W}$ to generate channel-wise context vectors.

$$F_{\text{avg}} = \frac{1}{H \times W} \sum_{i=1}^{H} \sum_{j=1}^{W} F_{i,j} \in \mathbb{R}^{C}. \tag{1}$$

$$F_{\max} = \max_{i,j} F_{i,j} \in \mathbb{R}^{C}. \tag{2}$$

where H represents height of the feature map. W is the width of feature map. C is the number of channels in the feature map. $F_{\text{avg}}^c \in \mathbb{R}^C$ is the channel level global average pooling feature vector, reflecting the global distribution characteristics of each channel. $F_{max}^c \in \mathbb{R}^C$ is the channel level max pooling feature vector, preserving the most significant activation response of each channel.

After processing through a shared Multi-Layer Perceptron (MLP), generate the channel weights s_c through Sigmoid activation (3).

$$s_c = \sigma(\text{MLP}(F_{\text{avg}}) + \text{MLP}(F_{\max})). \tag{3}$$

Then we get the output of the channel attention (4).

$$F_c = F \otimes s_c. \tag{4}$$

(2) Spatial Attention Calculation

Perform average pooling and maximum pooling on F_c along the channel dimension to generate spatial context maps.

$$F_{\text{avg}}^s = \frac{1}{C} \sum_{k=1}^{C} F_{c,k} \in \mathbb{R}^{H \times W} \tag{5}$$

$$F_{\max}^s = \max_{k} F_{c,k} \in \mathbb{R}^{H \times W} \tag{6}$$

After concatenation, generate the spatial weights s_s through 7×7 convolution and Sigmoid activation.

$$s_s = \sigma(\text{Conv}^{7 \times 7}([F_{\text{avg}}^s; F_{\max}^s])). \tag{7}$$

Then, we get the final output feature F_{attn}.

$$F_{\text{attn}} = F_c \otimes s_s \tag{8}$$

2.2.2 Embedding Position Optimization

Traditional CBAM is typically embedded directly after the convolutional layer. However, experiments have shown that this conventional embedding approach can lead to gradient vanishing in deep networks, particularly in small object detection tasks, where gradient backpropagation efficiency significantly deteriorates. This paper proposes a pre-activation strategy by adjusting the embedding position of the CBAM module and redesigning the activation function, substantially improving the gradient flow propagation path. The specific implementation is shown in Eq. (8).

$$F_{\text{out}} = \text{LeakyReLU}(\text{CBAM}(\text{BN}(\text{Conv}(F_{\text{in}})))) \tag{8}$$

The CBAM module is placed before the batch normalization (BN) layer, forming a cascaded structure of Conv → BN → CBAM → LeakyReLU. This design ensures that attention weights are applied to normalized feature maps, avoiding distribution bias caused by unstable convolutional output ranges.

This strategy expands the range of channel weights from [0,1] to [-1,1], enhancing the gradient propagation efficiency. Ablation experiments show that this strategy can effectively improve the recall rate of small target detection.

2.3 Improved BiFPN Feature Fusion

2.3.1 Bidirectional Feature Pyramid Structure

The improved BiFPN module introduces a sophisticated three-up and three-down bidirectional connectivity pattern, expanding upon the original BiFPN by incorporating an additional feature transmission pathway. This architectural enhancement significantly improves multi-scale feature fusion capabilities while maintaining computational efficiency. The modified structure operates through three distinct pathways.

(1) Top-Down Pathway

The high-level feature P_4 is fused with P_3 through upsampling as following (9).

$$P_3^{\text{up}} = \text{Upsample}(P_4) + P_3 \tag{10}$$

The upsampling operation uses learnable deconvolution kernels with stride = 2, and the summation implements element-wise feature aggregation. This pathway enables the propagation of rich semantic information from deeper layers to shallower levels, particularly beneficial for small object detection where high-level context is crucial.

(2) Bottom-Up Pathway

The low-level feature P_2 is fused with P_3^{up} through downsampling as following (10).

$$P_3^{\text{down}} = \text{Downsample}(P_2) + P_3^{\text{up}} \tag{11}$$

The downsampling convolution employs activation function and batch normalization to maintain feature stability. This bottom-up flow preserves fine-grained spatial details from early convolutional layers, which are essential for precise localization of traffic signs.

(3) Cross-Layer Skip Connection

By Introducing the lateral connection P_3^{skip}, and preserving the original features through residual connection, we get the feature P_3^{final}.

$$P_3^{final} = P_3^{down} + P_3^{skip} \qquad (12)$$

This skip connection effectively mitigates gradient vanishing by shortening backpropagation paths while preserving original feature details that might otherwise be suppressed during fusion. By maintaining feature diversity, it significantly enhances network stability. This integrated design ensures thorough propagation of deep-level features while preventing the loss of critical information caused by multiple feature transformations, thereby achieving more robust multi-scale feature representation in complex traffic scenarios.

2.3.2 Dynamic Weight Normalization

In response to the issue of gradient explosion that may occur in the weighted fusion of traditional BiFPN, this paper proposes Dynamic Weight Normalization (12). This method effectively resolves training stability problems in multi-scale feature fusion through an exponential normalization mechanism for learnable weights.

$$P_{out} = \sum_{i=1}^{n} \frac{\exp(w_i)}{\sum_{j=1}^{n} \exp(w_j)} P_i \qquad (13)$$

where w_i represents the learnable weight.

3 Experiments and Result Analysis

3.1 Datasets and Experimental Setup

The experiments in this paper were meticulously executed and validated by leveraging the publicly accessible Chinese Traffic Sign Dataset (CTSD)[25]. This dataset is specifically crafted to mirror the unique characteristics of Chinese road scenarios, making it highly relevant for research focused on traffic sign detection in the Chinese context. The CTSD is composed of a total of 1,100 images. A strategic division was made, allocating 700 images for the training phase and 400 for the testing phase. This split ensures a comprehensive evaluation of the model's performance during both the learning and validation stages. The image dimensions within the dataset predominantly fall into two categories: 1024 × 768 and 1280 × 720. Such specific dimensions are representative of common image resolutions in real - world traffic surveillance and related applications. The traffic signs in the CTSD are systematically categorized into four main classes. Among them, there are 264 red circular prohibitory signs, which are designed to convey restrictions and prohibitions to road users. Additionally, there are 129 yellow triangular warning signs, crucial for alerting drivers to potential hazards ahead. The dataset also includes 139 blue circular mandatory signs, which enforce specific actions or regulations. Besides these three major classes, there is a subset of other signs.

The experiment was conducted on a Windows based PC. Its core configuration is Intel Core i7-11700F CPU, with a running frequency of 2.5 GHz, which can efficiently handle various computing tasks in experiments and is the key to data processing and algorithm execution. The system is equipped with 16 GB of DDR3 memory, providing space for the storage and fast reading of large-scale datasets in the experiment, ensuring smooth data transmission, and avoiding the impact of memory bottlenecks on the experimental process. In addition, the NVIDIA GTX 3060 graphics card comes with 12 GB of video memory, which is crucial for parallel computing tasks such as deep learning. That can achieve fast data transmission and processing, greatly accelerating model training and inference.

3.2 Detection Performance

To comprehensively validate the effectiveness and superiority of the proposed method for the traffic sign detection task, this section presents an analysis from three aspects: visual detection results, quantitative performance metrics, and the impact of model architecture.

Fig. 2. Example of traffic sign detection results of the proposed method

Figure 2 demonstrates the detection effects of the proposed method under different traffic scenarios. Whether in complex road environments or situations with dense or sparse vehicles, the algorithm can accurately identify traffic signs and clearly label their positions and categories. From close - range to relatively long - distance, and from simple backgrounds to complex scenes, the sign detection is accurate and stable, reflecting the good robustness and adaptability of the algorithm. This effectively verifies that the algorithm in this paper is more superior and reliable than other methods in traffic - sign detection.

Through the intuitive detection image examples, the effectiveness of the algorithm in traffic - sign detection has been initially demonstrated. To more comprehensively evaluate the algorithm's performance, detailed training and testing were carried out, with the specific results shown in Table 1 and Table 2.

To comprehensively evaluate the performance of the model, we adopt the following key metrics.

Precision, also known as the positive predictive value, reflects the proportion of truly positive samples among all the samples predicted as positive by the model. Its calculation formula is defined in (13),

$$precision = \frac{TP}{TP + FP} \qquad (14)$$

where TP represents true positives, that is, the number of samples correctly classified as positive. FP represents false positives, that is, the number of samples misclassified as positive. The higher the precision, the stronger the reliability that the samples predicted as positive by the model actually belong to the positive class.

Recall, also known as the sensitivity or true positive rate, is used to measure the proportion of positive samples that the model can correctly identify among all the actual positive samples. The calculation formula is

$$Recall = \frac{TP}{TP + NF} \qquad (15)$$

where NF is false negatives, that is, the number of positive samples misclassified as negative. The higher the recall, the stronger the model's ability to capture positive samples, and the fewer positive samples are missed.

Mean Average Precision (mAP) is a crucial metric for evaluating the performance of object detection models. mAP@.5 refers to the mean average precision calculated when the Intersection over Union (IoU) threshold between the predicted bounding box and the ground-truth bounding box is set to 0.5. It first calculates the AP (Average Precision, obtained by computing the area under the precision - recall curve) for each class, and then takes the average of the APs of all classes. While mAP@ is usually calculated by averaging the APs at a series of IoU thresholds (such as from 0.5 to 0.95, with an interval of 0.05). mAP@.5 focuses on examining the detection ability of the model under a specific IoU standard, and mAP@ more comprehensively and comprehensively reflects the model's accuracy in localizing and classifying objects under different IoU requirements.

Table 1. Training results on CTSD

Class	Precision	Recall	mAP@.5	mAP@
all	0.921	0.942	0.968	0.824
danger	0.945	0.938	0.986	0.831
prohibitory	0.925	0.939	0.96	0.837
mandatory	0.893	0.95	0.957	0.804

Based on the comprehensive analysis of the training and testing results, the proposed model demonstrates overall stability and excellent performance. For the all category

Table 2. Testing results on CTSD

Class	Precision	Recall	mAP@.5	mAP@
all	0.925	0.939	0.968	0.823
danger	0.952	0.929	0.986	0.833
prohibitory	0.925	0.937	0.96	0.84
mandatory	0.896	0.95	0.958	0.796

in the overall data, the precision and mAP@.5 in the training and testing phases are nearly identical, while there are slight fluctuations in the recall and mAP@.5:.95, but the differences are minimal. This indicates that the model has strong generalization ability and performs stably on both the training and testing datasets.

Specifically, for each individual category, the precision of the danger category increased to 0.952 during testing, while the recall slightly decreased. However, the mAP@.5 and mAP@.5:.95 values are quite good, suggesting that the model's recognition ability for this category was further enhanced during testing, although there are still a small number of missed detections. The indicators of the prohibitory category in the testing phase are similar to those in the training phase, indicating that the model's detection effect for this category is stable. The recall of the mandatory category remains at a relatively high level of 0.95, but the mAP@.5:.95 decreased to 0.796 during testing, reflecting fluctuations in the detection performance of this category under strict positioning standards. This may be due to the more complex nature of such objects in the testing dataset or differences in the sample distribution.

Figure 3 presents the confusion matrix detailing the experimental results obtained using the CBAM-BiFPN model. Remarkably, the model showcases outstanding proficiency in identifying the danger class, achieving a flawless accuracy rate of 100% without any misclassifications. When it comes to the prohibitory class, the model exhibits strong recognition capabilities, with an accuracy of 98%. However, a small fraction (2%) of samples are incorrectly classified as background, resulting in false negatives (FN). Additionally, the model achieves perfect performance with a 100% accuracy rate in detecting the mandatory class. In summary, the CBAM-BiFPN model demonstrates high reliability in detecting danger, prohibitory, and mandatory traffic sign classes. To further enhance the model's overall detection performance, future research efforts should focus on refining the model architecture or training methodology, with a specific emphasis on reducing background false positive errors.

The precision curves for the YOLOv5 traffic sign detection experiment based on CBAM – BiFPN are shown in Fig. 3. In the low-confidence range, the precision for the danger, prohibitory, and mandatory traffic sign categories rapidly increases. As the confidence level continues to rise, the precision gradually approaches 1. The model maintains extremely high accuracy across all traffic sign categories, demonstrating excellent overall performance. Specifically, when the confidence level for the danger category reaches 0.96, the average precision achieves 1.00.

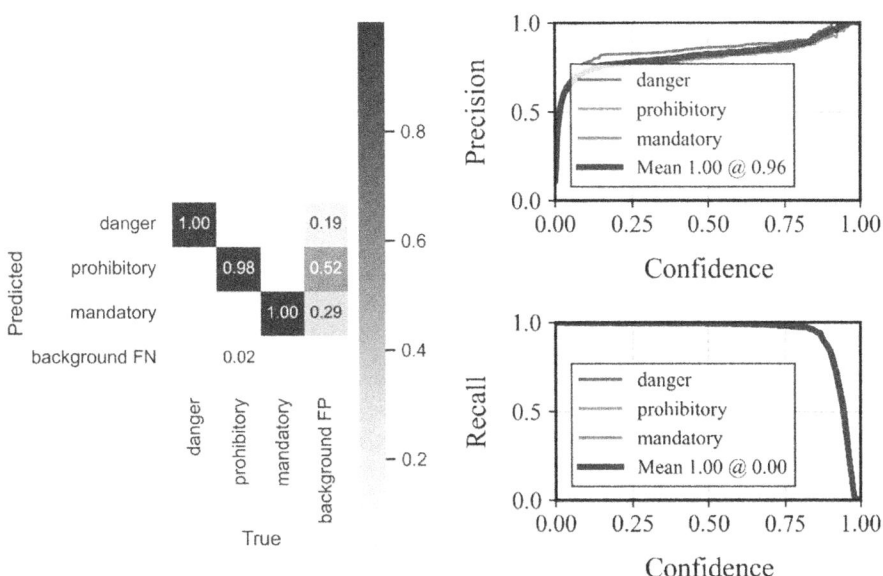

Fig. 3. Confusion matrix, Precision curves and Recall curves of traffic sign detection results of the proposed method

Figure 3 also depicts the recall-confidence curves for the YOLOv5 traffic sign detection experiment based on CBAM - BiFPN. In the low-confidence range, the recall rates are nearly 1, indicating that the model can identify most of the true traffic sign samples corresponding to each category during this phase. Moreover, as the confidence level increases, the recall rates remain high for an extended period, only dropping rapidly at high confidence levels.

3.3 Ablation Experiments

To comprehensively evaluate the effectiveness and performance advantages of the proposed algorithm in this paper, a series of ablation experiments were conducted for the object detection task. The experiments compared the performance of the baseline model and models with different improved modules introduced, encompassing key indicators such as precision, inference speed, model parameter count, and number of network layers. Detailed experimental results are presented in Table 3.

The ablation experiment results show that each model has its own advantages and disadvantages in terms of precision, speed, and complexity. The Baseline (YOLOv5s), as the basic model, has a precision of 90.4% and an FPS of 79. After adding CBAM, the speed significantly increases to 140 FPS, but the precision drops to 89%. When BiFPN is incorporated, the precision is 89.4% and the speed is 129 FPS. The proposed model performs the best, with a precision of 92.5%. Although its speed of 119 FPS is not the fastest and it has the most parameters (7.69M) and layers (232), it excels in balancing precision and speed, which proves the effectiveness and superiority of the proposed algorithm.

Table 3. Performance Ablation Results of Different Model Improvement Methods

Models	precision	FPS	Params (M)	layers
Baseline (YOLOv5s)	90.4%	79	7.02	213
+ CBAM	89%	140	7.54	224
+ BiFPN	89.4%	129	7.16	221
Ours	92.5%	119	7.69	232

4 Conclusion

This paper presents an enhanced YOLOv5 model by fusing the Convolutional Block Attention Module (CBAM) and Bidirectional Feature Pyramid Network (BiFPN), aiming to boost traffic sign detection performance in intricate scenarios. Traditional YOLOv5 models typically encounter difficulties with dynamic occlusion, extreme lighting, and multi-scale object detection. To surmount these obstacles, this study combines CBAM and BiFPN to devise an innovative architecture that dynamically strengthens key features and efficiently merges cross-scale information. Experimental results on the CTSD dataset show that the improved model attains an average detection accuracy (mAP@0.5) of 92.5%, marking a 2.1% increase compared to the original YOLOv5. Additionally, it retains a real-time processing speed of 119 frames per second (FPS), fulfilling the millisecond-level response demands of vehicle systems. Ablation experiments further confirm the effectiveness of the collaborative optimization of these modules.

The model's core innovations manifest in multiple aspects. CBAM modules are implanted at critical nodes in the backbone network. Through hierarchical and spatial attention operations, the model's capacity to concentrate on the main areas of traffic signs is notably enhanced. Meanwhile, the feature pyramid structure is transformed into BiFPN. By utilizing bidirectional cross-scale connections and a learnable weight normalization mechanism, it resolves the problem of shallow detail information attenuation prevalent in traditional Path Aggregation Networks (PANets). As a result, the enhanced model demonstrates greater robustness in complex scenarios, particularly when detecting distant traffic signs. Looking to the future, to address the ongoing challenges of dynamic occlusion, extreme lighting, and multimodal perception in complex traffic environments, a collaborative optimization architecture integrating deformable convolutions and Transformer will be developed. This novel architecture seeks to achieve a more profound integration of geometric perception and global context modeling.

Acknowledgments. This work is supported Xianyang City Key Research and Development Program Project (S2021ZDYF-SF-0739), University Youth Outstanding Talents Program of Shaanxi Province, Provincial-level Innovation and Entrepreneurship Training Program for University Students (S202410722038), National-level Innovation and Entrepreneurship Training Program for University Students (202410722004).

References

1. Taxonomy and Definitions for Terms Related to Driving Automation Systems for On-Road Motor Vehicles. SAE J3016_201609 (2016)
2. Sanyal, B., Mohapatra, R.K., Dash, R.: Traffic sign recognition: a survey. 2020 International Conference on Artificial Intelligence and Signal Processing, Amaravati, India, pp. 1–6 (2020)
3. Dalal, N., Triggs, B.: Histograms of oriented gradients for human detection. IEEE Computer Society Conference on Computer Vision and Pattern Recognition, IEEE, pp. 886–893 (2005)
4. Lowe, D.G.: Distinctive image features from scale-invariant keypoints. Int. J. Comput. Vision **60**(2), 91–110 (2004)
5. Cui, Y., Han, Y., Guo, D.: TS-DETR: multi-scale DETR for traffic sign detection and recognition. Pattern Recogn. Lett. **190**, 147–152 (2025)
6. Flores Calero, M.J., Aldas Sanchez, M., Vargas, J., Ayala, M.J.: Ecuadorian regulatory traffic sign detection by using HOG features and ELM classifier. IEEE Latin America Transactions **19**(4), 634–642 (2021)
7. Tang, J., Su, Q., Lin, C., Wen, Y., Su, B., Yang, J.: Traffic sign recognition based on HOG feature and SVM. 4th International Conference on Electronic Information Technology and Computer Engineering. New York, NY, USA: Association for Computing Machinery, pp. 534–538 (2021)
8. Ren, S., He, K., Girshick, R., Sun, J.: Faster R-CNN: towards real-time object detection with region proposal networks. IEEE Trans. Pattern Anal. Mach. Intell. **39**(6), 1137–1149 (2017)
9. Redmon, J., Divvala, S., Girshick, R., Farhadi, A.: You only look once: unified, real-time object detection. IEEE Conference on Computer Vision and Pattern Recognition, pp. 779–788 (2016)
10. Doherty, J., Gardiner, B., Kerr, E., Siddique, N.: BiFPN-YOLO: one-stage object detection integrating bi-directional feature pyramid networks. Pattern Recogn. **160**, 111209 (2025)
11. Liu, W., et al.: SSD: single shot multibox detector. Computer Vision – ECCV 2016, Springer International Publishing, Cham , pp. 21–37 (2016)
12. Xu, J.-H., Li, J.-P., Zhou, Z.-R., Lv, Q., Luo, J.: A survey of the yolo series of object detection algorithms. International Computer Conference on Wavelet Active Media Technology and Information Processing, pp. 1–6 (2024)
13. Hussain, M.: YOLOv1 to v8: unveiling each variant–a comprehensive review of YOLO. IEEE Access **12**, 42816–42833 (2024)
14. Mulajkar, R., Yede, S.: YOLO Version v1 to v8 comprehensive review. International Conference on Inventive Computation Technologies, pp. 472–478 (2024)
15. Redmon, J., Farhadi, A.: YOLO9000: better, faster, stronger. IEEE Conference on Computer Vision and Pattern Recognition, pp. 6517–6525 (2017)
16. Redmon, J., Farhadi, A.: YOLOv3: An Incremental Improvement (2018). https://doi.org/10.48550/arXiv.1804.02767
17. Tan, M., Pang, R., Le, Q.V.: EfficientDet: scalable and efficient object detection. IEEE Conference on Computer Vision and Pattern Recognition, pp. 10778–10787 (2020)
18. Ultralytics. Yolov5: A state-of-the-art real-time object detection system. Ultralytics Documentation (2021)
19. Dewi, C., Chen, R.-C., Liu, Y.-T., Jiang, X., Hartomo, K.D.: Yolo V4 for advanced traffic sign recognition with synthetic training data generated by various GAN. IEEE Access **9**, 97228–97242 (2021)
20. Liu, Y., Shi, G., Li, Y., Zhao, Z.: M-YOLO: traffic sign detection algorithm applicable to complex scenarios. Symmetry **14**(5), 952 (2022)
21. Wang, Y., Liu, Y., Yi, R., Jiang, Y.: Real-time traffic object detection algorithm with deep stochastic configuration networks. Inf. Sci. **700**, 121848 (2025)

22. Hou, Y., Zhang, Z., Du, L., Yin, J.: A fully locally selective large kernel network for traffic video detection. Measurement **242**, 115779 (2025)
23. Chen, Y., Luo, H.: VisioSignNet: a dual-interactive neural network for enhanced traffic sign detection. Expert Syst. Appl. **255**, 124688 (2024)
24. Wei, W., et al.: A lightweight network for traffic sign recognition based on multi-scale feature and attention mechanism. Heliyon **10**(4), e26182 (2024)
25. Yang, Y., Luo, H., Xu, H., Wu, F.: Towards real-time traffic sign detection and classification. IEEE Trans. Intell. Transp. Syst.Intell. Transp. Syst. **17**(7), 2022–2031 (2016)

Biomedical and Educational Data Science

Epidemiological and Transcriptomic Analysis of a Multidrug-Resistant *Pseudomonas Aeruginosa* Strain

Yao Zhou, Runqing Shi, Mengshan Zhou, Yuting Zhang, Shuo Wang, Yan Song, and Yaodong Chen(✉)

College of Life Sciences, Northwestern University, Xi'an 710069, Shaanxi, China
ydchen@nwu.edu.cn

Abstract. This study investigated the epidemiological characteristics, resistance profiles, and virulence factors of 102 multidrug-resistant *Pseudomonas aeruginosa* strains isolated from intensive care units (ICUs). Notably, the extensively drug-resistant strain X13 was identified and selected for in-depth transcriptomic analysis to explore its resistance mechanisms. Among the clinical isolates, carbapenem resistance was the most prevalent phenotype, while amikacin resistance remained low. Molecular typing revealed high genetic diversity, with ST1639 being the most common. Most strains harbored *exoT*, suggesting a conserved virulence profile.

Strain X13 was classified as ST244 and exhibited resistance to all tested antibiotics. Transcriptomic analysis revealed significant alterations in the expression of genes involved in cell wall synthesis, DNA repair, and stress responses, indicating their role in antibiotic resistance. The identification of differentially expressed genes and regulatory sRNAs (small RNAs) provides insight into the complex resistance mechanisms of X13. These findings contribute to a deeper understanding of multidrug resistance in *P. aeruginosa* and may inform future therapeutic strategies.

Keywords: *Pseudomonas aeruginosa* · Drug resistance · MLST · Virulence Factors · RNA-seq

1 Introduction

Pseudomonas aeruginosa is a common opportunistic pathogen in the Pseudomonas family. When a person is infected with *P. aeruginosa*, multiple infections may occur, including bloodstream infections (e.g., septicemia), as well as ocular, pulmonary, and urinary system infections [1, 2]. *P. aeruginosa* is present in the external environment and generally does not pose a threat to healthy individuals. However, when the immune system is compromised, the likelihood of infection significantly increases. As a member of the Pseudomonas genus, *P. aeruginosa* secretes virulence factors such as pyocyanin, fluorescein, and pyoverdine of different colors. Among them, pyocyanin appears blue-green under the microscope and is the key pathogenic factor causing hospital-acquired infections [3]. *P. aeruginosa* has a strong ability to adhere to the surface of contact lenses, on which it can form biofilms that are difficult to eliminate. If contact lenses are not thoroughly cleaned, there is a high risk of corneal infection, which can lead to blindness in severe cases [4, 5]. *P. aeruginosa* primarily infects the lungs of cystic fibrosis patients, leading to lung lesions. When the host's immune system is weakened, it is easy to cause *P. aeruginosa* infection, leading to skin and mucosal damage, and ultimately triggering dermatitis. At the same time, it is also the main pathogen causing burn wound infections [6–8]. *P. aeruginosa* has a large genome and a strong genetic reservoir capacity, allowing it to adapt to the environment through complex and diverse intracellular regulatory mechanisms. It can utilize different intracellular metabolic and virulence systems to enhance antibiotic resistance [9]. This widespread antibiotic resistance poses a significant challenge to current clinical treatments, making it increasingly difficult to treat diseases caused by this pathogen.

Multidrug-resistant *P. aeruginosa* is commonly encountered in hospitals around the world [10]. Due to its unique biological characteristics, *P. aeruginosa* exhibits strong drug resistance, which has become a major problem in clinical treatment. For *P. aeruginosa*, there are three primary antibiotic resistance mechanisms: natural resistance, acquired resistance, and adaptive resistance [11, 12]. Natural drug resistance is mainly related to decreased outer membrane permeability and expression of efflux pumps. Acquired resistance is achieved through mutations in resistance genes and the acquisition of exogenous resistance genes, particularly resistance to commonly used antibiotics such as β-lactams, aminoglycosides, and quinolones. This process is stable, heritable, and not affected by environmental changes. Obtaining exogenous resistance genes is achieved through the transfer of mobile genetic elements such as integrons, plasmids, and transposons [13, 14]. Unlike the other two mechanisms, adaptive resistance is dynamic and fluctuates with environmental changes. When the dosage of antimicrobial drugs or antibiotics changes, bacterial resistance may also change accordingly [15]. Bacterial adaptive resistance is reflected in the relative growth rate of bacteria in the host or environment, as well as their inter-host transmission capability.

A variety of techniques are used to analyze the characteristics and drug resistance of clinical bacteria. Multilocus sequence typing (MLST) uses DNA sequences of multiple housekeeping gene fragments to characterize isolated microorganisms, which is of significance for the prevention and control of drug resistance and pathogen spread, as well as for the molecular epidemiology of pathogens. Maiden et al. [16] were the first to apply the MLST typing method to *Neisseria meningitidis*. Since then, it has been widely adopted for typing various pathogens. This method is time-efficient and highly reproducible [17]. By selecting seven housekeeping genes, sufficient genetic diversity is ensured to differentiate closely related strains, enabling precise identification of the pathogen's genotype. This study employed MLST to characterize *P. aeruginosa* isolates from clinical samples and compared them with data from a database containing many clinical and environmental *P. aeruginosa* strains collected from different areas.

RNA sequencing (RNA-seq) has revolutionized the study of biology, particularly in the investigation of clinical drug-resistant bacteria. This technique plays a crucial role in mapping gene expression in drug-resistant bacteria under various physiological conditions, helping to elucidate the relationship between drug resistance and bacterial gene expression regulation [18, 19]. RNA-seq can be used to study the gene expression changes of antibiotic-resistant strains under the action of antibiotics. By analyzing differentially expressed genes, we aim to understand the transcriptional regulatory mechanisms of drug-resistant strains in response to antibiotic stress, thereby revealing the process of drug resistance formation [20, 21].

Although the genetic diversity and resistance patterns of *P. aeruginosa* have been extensively studied, the molecular mechanisms underlying pan-drug resistance in specific clinical isolates remain poorly understood. Strain X13, which exhibited resistance to all commonly used antibiotics, represents a rare and clinically significant case. However, the transcriptional adaptations and regulatory networks contributing to its extreme resistance phenotype have not been fully elucidated.

To address this gap, this study aimed to characterize the genomic and transcriptomic features of strain X13 in comparison to the wild-type PAO1, with the goal of identifying key genes and pathways involved in its resistance. We hypothesized that X13 exhibits distinct transcriptional signatures involving stress response, membrane transport, and resistance gene regulation that collectively contribute to its high-level resistance. By integrating MLST, antibiotic susceptibility profiling, and RNA-seq analysis, this study seeks to provide new insights into the resistance mechanisms of highly drug-resistant *P. aeruginosa* and inform potential therapeutic strategies.

2 Materials and Methods

2.1 Collection of *P. Aeruginosa* Clinical Strains and DNA Extraction

A collection of 102 antibiotic-resistant *P. aeruginosa* clinical isolates was gathered from Xi'an Children's Hospital in Shaanxi Province and Xi'an international medical center hospital. The strains were cultured from the clinical specimens, followed by bacterial identification. These isolates represented a variety of clinical specimen types. PAO1, maintained in our laboratory, was used as the wild-type reference. Genomic DNA was extracted from the strains using the Sangon Rapid Bacterial Genomic DNA Isolation Kit (Sangon Biotech, China). The DNA was then eluted in 50 μL of nuclease-free water and stored at $-80\ °C$ for future analysis.

2.2 Antimicrobial Susceptibility Testing

The antimicrobial susceptibility testing procedure involves incorporating antibiotics into Mueller-Hinton agar and monitoring bacterial growth, followed by the disk diffusion method for testing clinical isolates. Results are then interpreted based on established clinical guidelines [11, 22]. The antimicrobial susceptibility testing for the chosen strains were conducted through the disk diffusion method, and the results were interpreted following the guidelines provided by the Clinical and Laboratory Standards Institute. The antibiotics used include ciprofloxacin, tobramycin, cefepime, aztreonam, polymyxin, piperacillin, meropenem, imipenem, cefepime, ceftazidime, levofloxacin, gentamicin, and amikacin.

2.3 Determination of MLST and Virulence Factor

MLST analysis of *P. aeruginosa* involves identifying sequence variations in housekeeping genes. PCR amplicons are sent to Shanghai Sangon Biotech for sequencing, and the results are uploaded to the *P. aeruginosa* MLST database (https://pubmlst.org/paeruginosa/) for further analysis. Strains that do not match existing entries are categorized as new sequence types (ST). Virulence factors are evaluated by examining genes such as *plcH*, *aprA*, *algD*, *exoS*, *exoT*, *exoU*, *exoY*, *toxA*, and *nor*.

2.4 RNA-seq Technique

PAO1 and multidrug-resistant *P.aeruginosa* clinical strain X13 were inoculated onto LB agar and cultured for 8–12 h. A single colony was selected and cultured overnight in liquid LB medium to an OD_{600} of 0.6. After centrifugation, the sample was sent to Shanghai Meiji Biopharmaceutical Technology Co., Ltd. Gene expression quantification was performed using RSEM (RNA-Seq by Expectation-Maximization) with TPM (Transcripts Per Million) as the metric.

Differential gene expression analysis was performed using DESeq2 and edgeR, both of which employ negative binomial models. Genes with an adjusted p-value (p-adjust) < 0.05 and |log$_2$(fold change)| ≥ 1 were defined as significantly differentially expressed genes (DEGs). The p-adjust values were calculated using the Benjamini-Hochberg (BH) method for multiple testing correction, with p-adjust < 0.05 indicating statistical significance (Eq. 1), where m is the total number of tested genes, and rank(p_i) denotes the ascending rank of the raw p-value for gene i. The log$_2$(fold change) represents the base-2 logarithm of the ratio of expression levels between compared groups. The expression level Y_i of each gene follows a negative binomial distribution (Eq. 2), where μ_i is the expected count, and ϕ is the dispersion parameter accounting for overdispersion in count data [23].

$$\text{padji} = \frac{\text{pi} \cdot \text{m}}{\text{rank}(\text{pi})} \quad (1)$$

$$Y_i \sim \text{Negative Binomial}(\mu_i, \varphi) \quad (2)$$

DESeq2 begins its analysis with a read count matrix Y, where each row corresponds to a gene i and each column to a sample j. The matrix entry Y_{ij} represents the number of sequencing reads unambiguously mapped to gene i in sample j. DESeq2 uses the average normalized expression value of each gene across all samples as a filtering criterion, removing genes with mean normalized counts below a predefined threshold prior to multiple testing adjustment [24]. A generalized linear model (GLM) is then fitted, modeling the read counts Y_{ij} as following a negative binomial distribution. The mean μ_{ij} is defined as proportional to the true concentration of cDNA fragments from gene i in sample j, scaled by the sample-specific normalization factor s_j (Eq. 3). The normalization factor s_j for each sample is calculated using the Median-of-Ratios method. For each gene i, compute its geometric mean G_i across all samples, where Y_{ij} is the raw count of gene i in sample j, and m is the total number of samples (Eq. 4). For each sample j, the normalization factor s_j is defined as the median of the ratios of each gene's raw count Y_{ij} to its geometric mean G_i (Eq. 5). Next, for each gene i, compute its average normalized expression K_i (Eq. 6). Genes with K_i below a predefined threshold (default: 1) are filtered out.

$$\mu ij = sj \cdot qij \tag{3}$$

$$Gi = \left(\prod_{j=1}^{m} Yij\right)^{1/m} \tag{4}$$

$$sj = mediani\left(\frac{Yij}{Gi}\right) \tag{5}$$

$$Ki = \frac{1}{m}\sum_{j=1}^{m}\frac{Yij}{sj} \tag{6}$$

In edgeR, low-expression genes were filtered if their CPM (Counts Per Million) values were less than 1 across all samples (Eq. 7), where Y_{ij} represents the raw count of gene i in sample j, and N_j denotes the total sequencing depth (library size) of sample j [25]. A sample with the median library size (referred to as ref) was selected as the reference, and the TMM (Trimmed Mean of M-values) method was applied to calculate the normalization factor s_j for each sample. For each gene i, the log-fold change (M-value) (Eq. 8) and average expression (A-value) (Eq. 9) relative to the reference sample were computed. Genes with extreme M-values (top 30% and bottom 30% of the M-value distribution) were excluded to minimize the influence of highly expressed or differentially expressed genes on normalization. The weighted average of the M-values for the remaining 40% of genes was then calculated (Eqs. 10), yielding the normalization factor s_j for sample j (Eq. 11).

$$CPMij = \frac{Yij+0.5}{Nj/10^6} \tag{7}$$

$$Mi = \log_2\left(\frac{Yij/Nj}{Yi,ref/Nref}\right) \tag{8}$$

$$Ai = \frac{1}{2}\log_2((Yij/Nj)\cdot(Yi,ref/Nref)) \tag{9}$$

$$\log_2(sj) = \frac{\sum wiMi}{\sum wi} \tag{10}$$

$$sj = 2^{\log_2(sj)} \tag{11}$$

GO Functions and KEGG Pathway Enrichment Analysis. Gene Ontology (GO) enrichment analysis was employed to annotate the functional categories of differentially expressed genes (DEGs) and identify significantly enriched GO terms, aiding in the determination of their primary biological roles [26]. The analysis utilized the Gene Ontology database. Significant GO terms were identified through a hypergeometric test with a threshold of q < 0.05, compared to the genomic background. Pathway enrichment analysis was performed using the Kyoto Encyclopedia of KEGG database to identify pathways associated with the DEGs. Pathways with an FDR-adjusted P value (q < 0.05) were considered significantly enriched.

The fundamental principle of GO enrichment and KEGG pathway analysis relies on the hypergeometric distribution. The core equation calculates the probability of observing at least k genes from a specific gene set (e.g., a GO term or KEGG pathway) in a differentially expressed gene list of size n, given that the background genome contains M genes belonging to that functional set. This probability is compared against random expectations to determine significant enrichment. The p-value for Fisher's exact test is derived from the right-tailed cumulative probability of the hypergeometric distribution.

sRNA molecule prediction and analysis. Small RNAs (sRNAs) in bacteria are typically non-coding RNAs (ncRNAs) ranging from 50–500 bp. They are primarily located in intergenic regions but may also be found in the 5' and 3' untranslated regions (UTRs) of coding genes. Since the first identification of transcriptional regulatory factors in Escherichia coli in the mid-to-late 20th century, sRNAs have been recognized as crucial regulatory elements in bacteria. sRNAs are crucial in a range of biological processes, such as bacterial transcriptional regulation, RNA processing and modification, mRNA stability, protein translation and degradation, plasmid replication, and bacterial pathogenesis[31]. In this study, Rockhopper software was utilized to compare and annotate the newly identified intergenic region transcripts against the nr database. The sRNA sequence alignment was performed using NCBI tools, and target gene predictions were made using the IntaRNA web service.

CARD and VFDB Annotations. The Drug Resistance Gene Database (CARD, https://card.mcmaster.ca/) and Virulence Factor Database (VFDB, https://cge.cbs.dtu.dk/services/Res Finder/) are widely used annotation databases in transcriptome sequencing data analysis. CARD is designed to identify bacterial resistance genes and their associated mechanisms, while VFDB focuses on identifying genes related to bacterial pathogenicity. By aligning RNA-seq data with these two databases, researchers can pinpoint bacterial resistance or pathogenicity genes, providing valuable insights for the prevention and treatment of bacterial infections.

Data statistics and analysis. Statistical was using DESeq2 used for small sample size data, based on the negative binomial distribution model, suitable for differential expression analysis of RNA-seq data, DEGs with a false discovery rate (FDR < 0.01) and a fold change of ≥ 2 were selected for analysis. To control for false positives due to multiple testing, p-values were adjusted using the Benjamini-Hochberg procedure, which controls the false discovery rate (FDR) by ranking the raw p-values and determining significance based on a FDR threshold (0.05).

3 Results

3.1 Analysis on the Sources of Clinical Strains of *P. Aeruginosa*

This research analyzed 102 *P. aeruginosa* isolates collected from clinical settings, with samples originating from various human tissues. Among these, 82 strains were isolated from sputum, which represented most of the samples (80.39%). The second most common source was urine, from which 6 strains were isolated, followed by 8 strains from blood and skin secretion samples. Additionally, 6 strains were isolated from catheter drainage fluid and bronchoalveolar lavage fluid (see Fig. 1A). Most *P. aeruginosa* strains were collected from the intensive care unit, totaling 72 strains (70.59%) (see Fig. 1B), which is consistent with similar findings both domestically and internationally.

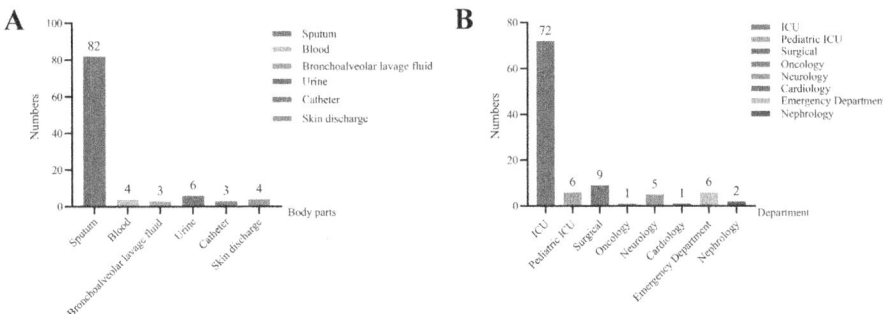

Fig. 1. Analysis of strain origin. The highest number of strains was isolated from sputum, with a total of 82 strains, followed by urine (A). Among all departments, the intensive care unit had the largest number of samples, with 72, accounting for 70.59% of the total (B).

3.2 Antimicrobial Susceptibility Analysis.

Table 1. Phenotypic analysis of 102 *P. aeruginosa* Isolates.

Types of antibiotics	Antibiotic	Number	Percentage
Aminoglycoside	Amikacin	3	2.94%
	Gentamicin	12	11.76%
	Tobramycin	6	5.88%
Quinolones	Ciprofloxacin	43	42.16%
	Levofloxacin	57	55.88%
Polypeptides	Ploymymin	0	0.00%
Monolactams	Aztreonam	47	46.08%
Cephalosporins	Ceftazidime	35	34.31%
Fourth-generation cephalosporin	Cefepime	38	37.25%

(*continued*)

Table 1. (*continued*)

Types of antibiotics	Antibiotic	Number	Percentage
Carbapenems	Meropenem	64	62.75%
	Imipenem	67	65.69%
ureidopenicillins	Piperacillin	42	41.18%
β-lactam	Piperacillin/tazobactam	39	38.24%
	Cefepime/sulbactam	35	34.31%

Antibiotic resistance represents one of the most significant global public health challenges. In this study, 102 clinical samples were tested for resistance to 14 commonly used antibiotics (Table 1). The results revealed that the resistance rate of P. aeruginosa infections in hospitals to aminoglycosides was low, with resistance rates to amikacin, gentamicin, and tobramycin at 2.94%, 11.76%, and 5.88%, respectively. However, the resistance rates to quinolone, ciprofloxacin and levofloxacin, were significantly higher at 42.16% and 55.88%, respectively. Resistance to aztreonam was 46.08%, and resistance to cephalosporins (ceftazidime and cefepime) ranged between 34.31% and 37.25%. Resistance to the combination of piperacillin/tazobactam and Cefepime/sulbactam also fell within the 30.00%–40.00%, there was no resistance to polymyxins. Most of the clinical samples collected were from multidrug-resistant bacteria. Therefore, to control the widespread infection of P. aeruginosa and the growth of multidrug-resistant strains in hospitals, it is crucial to provide patients with appropriate antibiotic treatment based on epidemiological data.

In this study, we examined the distribution of drug-resistant genes in 102 clinical *P. aeruginosa* samples, identifying 16 distinct resistance genes (see Fig. 2). The aminoglycoside resistance gene prevalence was as follows: *rmtB* (9/102, 8.82%) and *aac(6')Ib* (5/102, 4.90%). No *rmtA* or *armA* genes were detected, suggesting that *P. aeruginosa* aminoglycoside resistance is primarily mediated by the 16S rRNA methylase encoded by the *rmtB* gene. The carriage rates for quinolone resistance genes were *parC* (32/102, 31.37%) and *gyrA* (21/102, 20.59%), while *qnrA* and *qnrC* were absent. This implies that quinolone resistance in *P. aeruginosa* is predominantly linked to the topoisomerase *parC* gene and the DNA gyrase gene *gyrA*. The carriage rates of β-lactam resistance genes included *foxA* (35/102, 34.31%), bla_{OXA} (24/102, 23.53%), bla_{TEMP} (2/102, 1.96%), bla_{GES} (5/102, 4.90%), and bla_{SHV} (3/102, 2.94%). The bla_{DHA} and bla_{IMP} genes were not detected. Notably, 43 strains did not carry any of the aminoglycoside resistance genes screened (see Fig. 2). These results suggest that *P. aeruginosa* resistance to aminoglycosides is mainly attributed to chromosomal *fox* gene and plasmid-mediated bla_{OXA} gene.

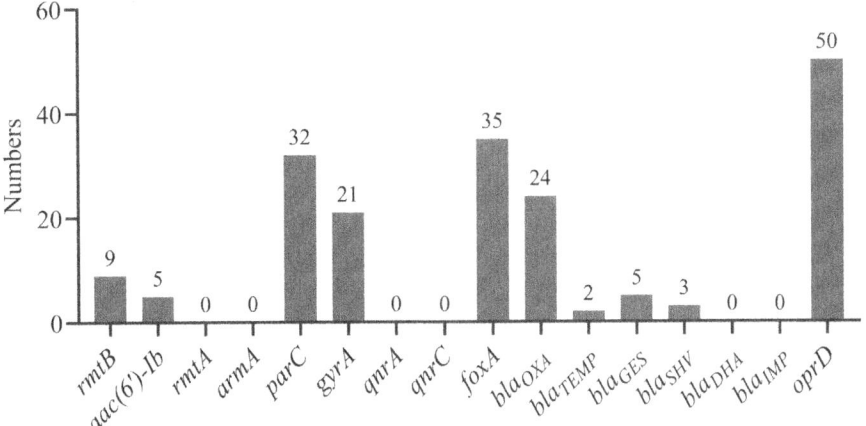

Fig. 2. Correlation analysis between drug-resistant genes and phenotypes of *P. aeruginosa* Isolates.

3.3 Molecular Characteristics of the *P. aeruginosa* Isolates

To explore the distribution and epidemiological characteristics of *P. aeruginosa*, MLST was employed for strain classification. MLST analysis of 102 P. aeruginosa strains revealed that ST1639 was the predominant genotype, followed by ST261, ST485, ST639, and ST277. A total of 45 distinct ST types were identified among the isolates. Notably, ST1639 accounted for 14.71% of the strains. Of particular interest was the identification of one *P. aeruginosa* strain, designated X13, which exhibited resistance to all tested antibiotics; this strain was classified as ST244. The statistical data is presented in Fig. 3A.

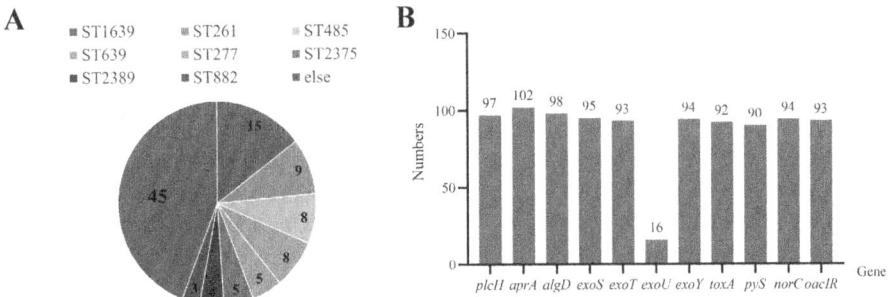

Fig. 3. MLST typing and Virulence Factors analysis of *P. aeruginosa* Isolates. A) MLST typing shows that ST1639 is the predominant genotype, followed by ST261, ST485, ST639, and ST277. Among the 102 isolated strains of *P. aeruginosa*, a total of 45 ST types were identified. B) Virulence Factors analysis shows that 93 strains (91.17%) carried the *exoT* gene, which is highly conserved in *P. aeruginosa* and plays a crucial role in its pathogenic mechanism. Additionally, 94 strains (92.16%) carried the *exoY* gene, which is commonly found in *P. aeruginosa* biofilms.

3.4 Virulence Factors Analysis the *P. aeruginosa* Isolates

The presence of virulence genes indicates the pathogenic potential of bacteria. In this study, PCR was employed to detect several virulence factors in clinical isolates. The *aprA* gene is associated with the type I secretion system (T1SS), while *toxA* and *plcH* are linked to the type II secretion system (T2SS). Furthermore, *exoS*, *exoT*, *exoY*, and *exoU* are related to the type III secretion system (T3SS).

The virulence genes carried by strains of the same MLST can vary, and no significant association was found between them. In this study, 102 *P. aeruginosa* strains isolated from clinical sites were tested (Fig. 3B), and the results indicated that T3SS-related virulence factors were present in almost all the pathogens. Of the isolates, 93 strains (91.17%) carried the *exoT* gene, which is highly conserved in *P. aeruginosa* and plays a crucial role in its pathogenic mechanism. Additionally, 94 strains (92.16%) carried the *exoY* gene, which is commonly found in *P. aeruginosa* biofilms. The formation of biofilms can lead to recurrent and difficult-to-treat infections. This gene also contributes to the rearrangement of the actin cytoskeleton. Furthermore, 95 strains (93.14%) carried the *exoS* gene, which is significant in respiratory infections and is frequently detected in conditions like pneumonia. Only 16 strains (15.69%) carried the *exoU* gene. Despite having the lowest carriage rate, Azimi et al. found that its cytotoxicity is 100 times greater than that of *exoS* [27]. The secretion of *exoU* is closely linked to pathogenicity and is strongly correlated with adverse clinical outcomes. In conclusion, virulence genes associated with T3SS are abundant in clinical isolates of *P. aeruginosa* from nosocomial infections, indicating that this system plays a crucial role in the pathogenic mechanism of the pathogen.

3.5 Epidemiological Analysis and RNA-Seq Analysis of X13 Isolate

Interestingly, in our previous analysis of drug resistance, we identified a strain resistant to antibiotics commonly used in hospitals, which we designated as strain X13. Except for showing intermediate resistance to polymyxin, it was resistant to all other tested antibiotics. Epidemiological testing revealed that the isolate was obtained from the sputum of a patient in the ICU. Its MLST was ST244, a less common type compared to ST1639 in the hospital where the samples were collected. Virulence gene testing showed that the strain harbored nearly all known virulence genes, providing a basis for further investigation of its toxicity. Subsequent studies focused on analyzing the strain's transcriptome to explore its genetic characteristics.

GO functions and KEGG pathway enrichment analysis. X13 was selected for RNA-seq because it displayed the most extensive resistance phenotype, including resistance to cephalosporins, carbapenems, and aminoglycosides, and harbored multiple resistance genes such as bla_{VEB} and bla_{OXA-10}. We focused on this strain to better understand the transcriptional basis of its multidrug resistance, as representative of the most challenging clinical isolates.

Three replicate samples were collected from the wild-type PAO1 and the experimental group X13, respectively. RNA was subsequently extracted, followed by transcriptome sequencing. The Raw Error Rate (%) reflects the error rate in the original sequencing data, with a lower percentage indicating better data quality. Figure 4 reflects the extremely low error rate of the wild-type PAO1 and experimental group X13. Raw Q20 (%) represents the percentage of bases with a quality score greater than or equal to 20 in the original data, with a higher Q20 value indicating better data quality. The results demonstrated that the quality of all samples was very high, with the Q20 values of the raw data ranging from 98.33% to 98.45%, and the Q30 values ranging from 95.28% to 95.50%. After quality control of the data, the Q20 values for all samples slightly increased (98.82% to 98.89%), and the Q30 values also showed an improvement, ranging from 96.14% to 96.29% (Table 2).

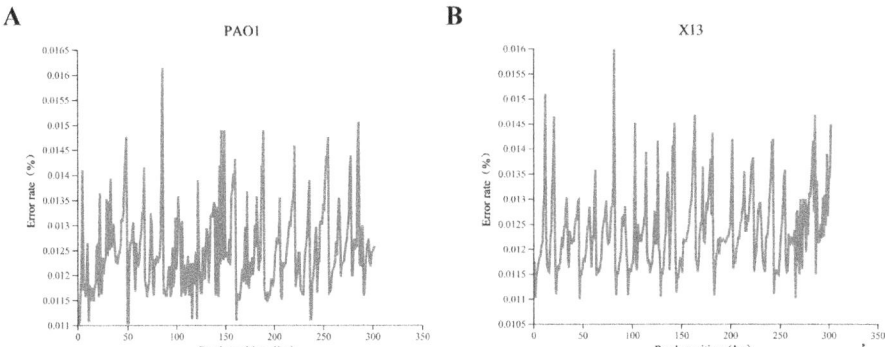

Fig. 4. Distribution of base error rates of raw reads of wild-type PAO1 (A) and experimental group X13 (B).

Table 2. Quality control data statistics table.

Sample Name	Raw Bases (bp)	Raw Error Rate (%)	Raw Q20 (%)	Raw Q30 (%)	Clean Q20 (%)	Clean Q30 (%)
X13-1	3502121558	0.0123	98.42	95.46	98.87	96.24
X13-2	3780058802	0.0124	98.37	95.35	98.85	96.19
X13-3	4434222512	0.0123	98.44	95.5	98.89	96.29
PAO1-1	3938550818	0.0125	98.33	95.28	98.82	96.14
PAO1-2	4328008206	0.0124	98.42	95.44	98.86	96.23
PAO1-3	3742671202	0.0123	98.45	95.45	98.83	96.14

Note: Raw reads: the number of paired-end reads in the original sequence data; (3) Raw bases: the total number of bases in the original data obtained by multiplying the number of raw paired-end reads by the length; (4) Raw Error Rate: the raw base error rate; (5) Raw Q20, Q30: the percentage of bases with Phred values greater than 20 and 30 in the total bases calculated from the raw data; (6) Clean Reads: the number of paired-end reads after quality control; (9) Clean Q20, Q30: the percentage of bases with Phred values greater than 20 and 30 in the total bases calculated from the quality control

Differential Gene Expression (DGE) Analysis. The differential analysis software edgeR can serve as an alternative tool for performing differential gene expression analysis. It allows for analysis with or without biological replication. The default thresholds are: adjusted p-value < 0.05 and |log2FC| ≥ 1. In this study, edgeR was used to identify differentially expressed genes (DEGs) in the sequencing data of the wild-type PAO1 and the experimental group X13. The screening criteria were p-value < 0.05 and |log2FC| > 1 (indicating a fold change greater than 2). Figure 5A presents a volcano plot of the differential gene expression between X13 and PAO1. Blue represents genes that are differentially upregulated, red represents genes that are differentially downregulated, and gray represents genes with no significant difference. As shown in the plot, 3,856 genes (gray dots) showed no significant differential expression. In comparison to the wild-type PAO1, 872 genes in X13 were upregulated at the transcriptional level, while 1,342 genes were downregulated. The results from DESeq2 were consistent with edgeR in identifying upregulated genes, whereas some differences were observed in the classification of downregulated and non-significant genes.

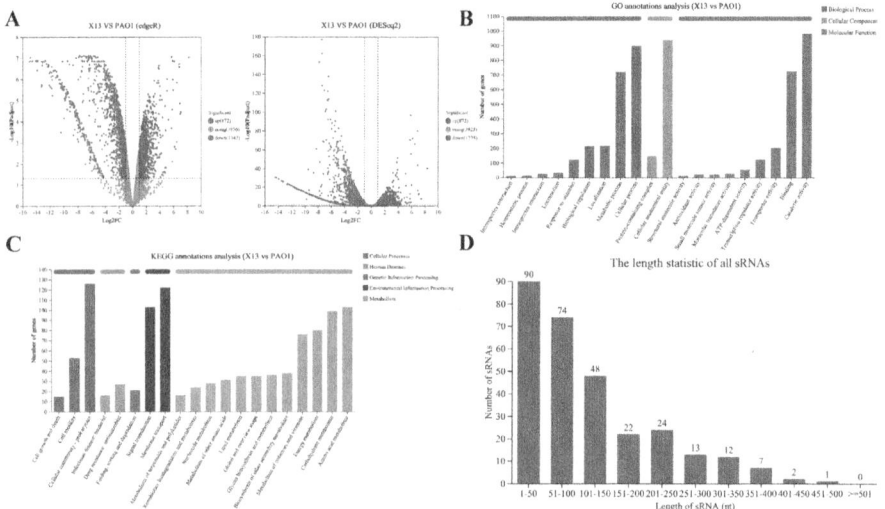

Fig. 5. RNA-seq analysis of *P. aeruginosa* X13 isolate. A. Volcano plots showing differential gene expression between Pseudomonas aeruginosa X13 and PAO1 analyzed by edgeR (left) and DESeq2 (right). B. X13-GO pathway enrichment analysis shows that within the biological process category, "cellular process" and "metabolic process" were predominant, accounting for 896 and 720 genes, respectively, while fewer genes were associated with intraspecific biological processes and homeostatic processes. C. X13-KEGG pathway enrichment analysis indicates that over 170 genes were linked to "cell growth and death" and "cell motility". D. Base length statistics of X13 candidate sRNAs reveals that most of the sRNAs had a base length within 350 nucleotides.

The differential analysis software edgeR can serve as an alternative tool for performing differential gene expression analysis. It allows for analysis with or without biological replication. The default thresholds are adjusted p-value < 0.05 and |log2FC| ≥ 1. In this study, edgeR was used to identify differentially expressed genes (DEGs) in the sequencing data of the wild-type PAO1 and the experimental group X13. The screening criteria were p-value < 0.05 and |log2FC| > 1 (indicating a fold change greater than 2). Figure 5A presents a volcano plot of the differential gene expression between X13 and PAO1. Blue represents genes that are differentially upregulated, red represents genes that are differentially downregulated, and gray represents genes with no significant difference. As shown in the plot, 3,856 genes (gray dots) showed no significant differential expression. In comparison to the wild-type PAO1, 872 genes in X13 were upregulated at the transcriptional level, while 1,342 genes were downregulated. The results from DESeq2 were consistent with edgeR in identifying upregulated genes, whereas some differences were observed in the classification of downregulated and non-significant genes.

GO Enrichment Analysis of DEGs. GO analysis is a comprehensive database established by the Gene Ontology Consortium, which categorizes and summarizes all research results related to genes worldwide. This database consists of three categories, namely biological process, cellular component, and molecular function, each describing the biological processes involved in gene products, the cellular environment in which they exist, and the molecular functions they may perform. The X13 strain has 2379 genes with GO annotation function, of which biological processes account for 41.24%, cellular components account for 20.00%, and molecular functions account for 38.76%. This indicates that the gene products of X13 are mainly concentrated in biological processes (Fig. 5B). In the biological process category, "cellular process" and "metabolic process" are dominant, with 896 and 720 genes, respectively. Fewer genes are associated with intraspecific biological processes and homeostatic processes. In the cellular component category, most genes are linked to "organelle structure," with only a few involved in protein complexes. In the molecular function category, "catalytic activity" and "binding activity" are the most common functions, with 982 and 725 genes, respectively, while the number of genes associated with "antioxidant activity" and "structural molecular activity" is considerably smaller. Overall, the genes are primarily concentrated in cellular processes, metabolic processes, catalytic activity, and binding activity, while antioxidant and structural functions are less prevalent.

KEGG Pathway Enrichment Analysis of DEGs. KEGG metabolic pathway analysis can provide valuable insights into the metabolic functions associated with genes, offering new directions for further experimental research. To explore the distribution of genes across different functional pathways and categories in X13, this study conducted KEGG Pathway enrichment analysis. The results are as follows (Fig. 5C).

The top 20 *P. aeruginosa* pathways involved in pathway enrichment mainly fall into four categories. In the category of cellular processes, "cellular community prokaryotes" has the largest proportion, with 126 genes associated with this pathway, followed by "cell motility", indicating that genes associated with these processes are highly active. In the category of Human Diseases, the main causes are bacterial infections and antibiotic resistance. Therefore, it can be inferred that for X13, drug resistance has a significant impact on the pathogenicity of bacteria. In the category of Genetic Information Processing, the number of genes involved in the "folding, sorting, and degradation" pathway is relatively small, indicating that these genes related to information transmission and gene expression regulation are not very active in this dataset. In the Environmental

Information Processing category, particularly the "membrane transport" pathway, more gene annotations are shown, emphasizing the important role of cell membrane response to environmental signals in this sample. Metabolism is the most prominent category in analysis, involving a much higher number of genes compared to other categories, highlighting a wide range of metabolic activities in the sample. Among them, the genes involved in carbohydrate metabolism and amino acid metabolism of *P. aeruginosa* are the most abundant, indicating their core importance in gene expression and cellular activity.

sRNA molecule prediction and target gene function analysis. Bacterial small RNAs (sRNAs) are a type of non-coding RNA, typically ranging in length from 50 to 500 nucleotides. They have a stable secondary structure and play crucial roles in regulating bacterial stress responses, host infection, quorum sensing, and other cellular activities [28, 29]. To date, research on the sRNAs of *P. aeruginosa* remains in its early stages, and the sRNAs and their target regulatory genes that have been predicted and identified so far represent only a small fraction of the bacterial sRNA library. Therefore, this study aimed to predict sRNAs based on transcriptome data. New intergenic region transcripts were identified and compared against the nr database using Rockhopper software. Transcripts that could not be annotated were considered potential non-coding sRNAs, resulting in the identification of 293 candidate sRNAs. The base sequences of these 293 sRNAs were then analyzed, revealing that most of the sRNAs had a base length within 350 nucleotides, with more than half falling within 100 nucleotides. The majority of sRNAs had lengths between 1 and 50 nucleotides. These results suggest that a significant portion of the non-coding sRNAs identified are relatively short, aligning with the typical length requirements for sRNAs (Fig. 5D).

CARD and VFDB Annotation Analysis. The analysis of drug resistance and virulence factors of X13 revealed some key findings. After annotating and categorizing the CARD data, it was found that efflux pumps accounted for the largest proportion, with resistance inhibition cell division (RND) and major facilitator superfamily (MFS) families being the main mechanisms of drug resistance. For antibiotic resistance genes, the proportion of cephamycin and cephamycin genes is relatively small (Fig. 6A). After annotating and categorizing VFDB data, it was found that nutritional/metabolic factors accounted for the largest proportion, with a total of 212 genes, indicating that these genes play important roles in bacteria. In addition, the proportions of effect delivery system (118 genes), adhesion (97 genes) and immune module (93 genes) are similar. Although the number of other virulence factors such as motility (64 genes) and biofilm (57 genes) are relatively small, they also play a crucial role in bacterial pathogenicity and resistance (Fig. 6B). These findings suggest that the overall profile of X13 reflects a complex set of resistance and virulence factors critical to its survival and pathogenicity.

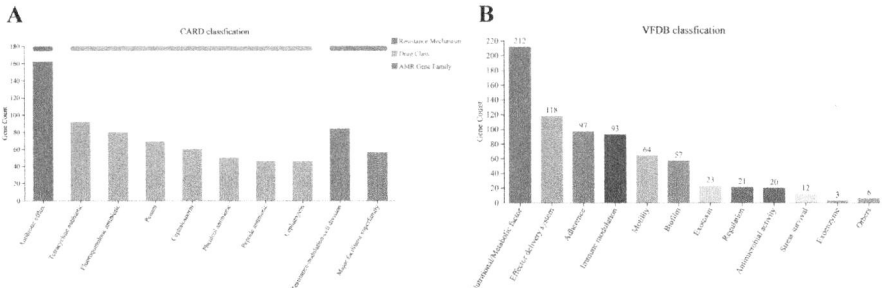

Fig. 6. CARD and VFDB annotation analysis of X13. A. CARD and VFDB annotation analysis show that the RND antibiotic efflux pump and antibiotic efflux pump categories dominated the gene count, B. VFDB annotation analysis shows that the Nutritional/Metabolic factor category had the highest number of genes, reaching 212.

4 Discussion

As an important opportunistic pathogen, the pathogenicity of *P. aeruginosa* is closely linked to its complex and highly regulated regulatory systems [30]. Epidemiological data from 102 clinical strains of *P. aeruginosa* revealed that all were multidrug-resistant, with the majority isolated from the intensive care unit (ICU) of the hospital. Over 80.00% of the samples were obtained from patient sputum. A total of 45 distinct MLST types were identified, with ST1639 being the most prevalent, followed by ST261, indicating that ST1639 is a common strain in hospitals in Xi'an. However, our results diverge from global reports, where ST111 and ST235 are often the most prevalent in European and North American settings, respectively. This discrepancy may be attributed to local epidemiological factors and the circulation of different clonal lineages in hospital environments [31]. *P. aeruginosa* expresses a variety of [32, 33], including flagella, pili, and pyocyanin. The expression levels, synergistic interactions, and regulation of these virulence factors are crucial in determining the pathogen's potential. Virulence factor analysis revealed that the detection rate of *aprA*, associated with the type I secretion system, was 100.00%. Additionally, *exoY* (which disrupts endothelial cell integrity), *exoS* (which induces rearrangement of the actin cytoskeleton), and *exoT* (which inhibits neutrophil division)—all linked to the type III secretion system—were highly expressed in *P. aeruginosa*.

The results of RNA-seq revealed that, compared to the wild type, X13 exhibited 856 upregulated and 1,236 downregulated genes at the transcriptional level, which aligns with previous global transcriptomic studies that identified similar adaptive gene expression patterns in *P. aeruginosa* under antimicrobial pressure [34]. Gene Ontology (GO) analysis showed that within the biological process category, "cellular process" and "metabolic process" were predominant, accounting for 896 and 720 genes, respectively, while fewer genes were associated with intraspecific biological processes and homeostatic processes. KEGG pathway analysis indicated that over 170 genes were linked to "cell growth and death" and "cell movement", emphasizing the significance of these cellular activities in the dataset. Furthermore, 293 non-coding sRNAs were identified in X13, which are crucial in regulating bacterial stress responses, quorum sensing, and other related processes.

Epidemiological analysis of *P. aeruginosa* is crucial for understanding its transmission patterns and providing data to support the formulation of public health strategies. Strain X13's unique resistance mechanisms, particularly its upregulation of efflux pumps and stress response genes, highlight the need for targeted antibiotic stewardship strategies to prevent further spread of multidrug-resistant strains in clinical settings. Overall, RNA-seq reveals changes in gene expression of *P. aeruginosa* across different environments and hosts, aiding in the identification of key pathogenic factors and mechanisms of drug resistance. The integration of these two approaches enhances our understanding of the bacteria's pathogenic mechanisms and provides a theoretical foundation for developing new diagnostic methods.

Funding. This research was funded by the National Natural Science Foundation of China (grant no. 31970050) to YC.

References

1. Blomquist, K.C., Nix, D.E.: A critical evaluation of newer beta-lactam antibiotics for treatment of pseudomonas aeruginosa infections. Ann. Pharmacother. **55**, 1010–1024 (2021)
2. Haque, M., Sartelli, M., McKimm, J., Abu Bakar, M.: Health care-associated infections - an overview. Infect Drug Resist **11**, 2321–2333 (2018)
3. Qin, S., et al.: Pseudomonas aeruginosa: pathogenesis, virulence factors, antibiotic resistance, interaction with host, technology advances and emerging therapeutics. Signal Transduct. Target. Ther. **7**, 199 (2022)
4. Diekema, D.J., et al.: Trends in antimicrobial susceptibility of bacterial pathogens isolated from patients with bloodstream infections in the USA, Canada and Latin America. SENTRY Participants Group. Int J Antimicrob Agents **13**, 257–271 (2000)
5. Li, J., et al.: Characterization of clinical extensively drug-resistant pseudomonas aeruginosa in the Hunan province of China. Ann. Clin. Microbiol. Antimicrob. **15**, 35 (2016)
6. Rossi, E., et al.: Pseudomonas aeruginosa adaptation and evolution in patients with cystic fibrosis. Nat. Rev. Microbiol. **19**, 331–342 (2021)
7. Pang, Z., Raudonis, R., Glick, B.R., Lin, T.J., Cheng, Z.: Antibiotic resistance in pseudomonas aeruginosa: mechanisms and alternative therapeutic strategies. Biotechnol. Adv. **37**, 177–192 (2019)
8. Rees, V.E., et al.: Characterization of hypermutator pseudomonas aeruginosa isolates from patients with cystic fibrosis in Australia. Antimicrob Agents Chemother **63**, e02538–18 (2019)
9. Chatterjee, M., Anju, C.P., Biswas, L., Anil Kumar, V., Gopi Mohan, C., Biswas, R.: Antibiotic resistance in pseudomonas aeruginosa and alternative therapeutic options. Int. J. Med. Microbiol. **306**, 48–58 (2016)
10. Aslan, A.T., Ezure, Y., Horcajada, J.P., Harris, P.N.A., Paterson, D.L.: In vitro, in vivo and clinical studies comparing the efficacy of ceftazidime-avibactam monotherapy with ceftazidime-avibactam-containing combination regimens against carbapenem-resistant Enterobacterales and multidrug-resistant Pseudomonas aeruginosa isolates or infections: a scoping review. Front Med (Lausanne) **10**, 1249030 (2023)
11. Denis, J.B., et al.: Multidrug-resistant pseudomonas aeruginosa and mortality in mechanically ventilated ICU patients. Am. J. Infect. Control **47**, 1059–1064 (2019)

12. Zheng, D., Bergen, P.J., Landersdorfer, C.B., Hirsch, E.B.: Differences in fosfomycin resistance mechanisms between pseudomonas aeruginosa and enterobacterales. Antimicrob. Agents Chemother. **66**, e0144621 (2022)
13. Khan, Z., Suthanthiran, M., Muthukumar, T.: MicroRNAs and transplantation. Clin. Lab. Med. **39**, 125–143 (2019)
14. Harrison, S.A., et al.: NIS2+™, an optimisation of the blood-based biomarker NIS4® technology for the detection of at-risk NASH: a prospective derivation and validation study. J. Hepatol. **79**, 758–767 (2023)
15. Fernandez-Sanchez, A., et al.: Inflammation, oxidative stress, and obesity. Int. J. Mol. Sci. **12**, 3117–3132 (2011)
16. Maiden, M.C., et al.: Multilocus sequence typing: a portable approach to the identification of clones within populations of pathogenic microorganisms. Proc Natl Acad Sci U S A **95**, 3140–3145 (1998)
17. Curran, B., Jonas, D., Grundmann, H., Pitt, T., Dowson, C.G.: Development of a multilocus sequence typing scheme for the opportunistic pathogen pseudomonas aeruginosa. J. Clin. Microbiol. **42**, 5644–5649 (2004)
18. Han, H.: Diagnostic biases in translational bioinformatics. BMC Med. Genomics **8**, 46 (2015)
19. Han, H., Jiang, X.: Disease biomarker query from RNA-Seq data. Cancer Inform **13**, 81–94 (2014)
20. Thoming, J.G., Haussler, S.: Transcriptional profiling of pseudomonas aeruginosa infections. Adv. Exp. Med. Biol. **1386**, 303–323 (2022)
21. Leinweber, A., Laffont, C., Lardi, M., Eberl, L., Pessi, G., Kummerli, R.: RNA-Seq reveals that pseudomonas aeruginosa mounts growth medium-dependent competitive responses when sensing diffusible cues from Burkholderia cenocepacia. Commun Biol **7**, 995 (2024)
22. Blair, J.M., Webber, M.A., Baylay, A.J., Ogbolu, D.O., Piddock, L.J.: Molecular mechanisms of antibiotic resistance. Nat. Rev. Microbiol. **13**, 42–51 (2015)
23. Love, M.I., Huber, W., Anders, S.: Moderated estimation of fold change and dispersion for RNA-seq data with DESeq2. Genome Biol. **15**, 550 (2014)
24. Maza, E.: In Papyro comparison of TMM (edgeR), RLE (DESeq2), and MRN normalization methods for a simple two-conditions-without-replicates RNA-Seq experimental design. Front Genet **7**, 164 (2016)
25. Robinson, M.D., McCarthy, D.J., Smyth, G.K.: EdgeR: a bioconductor package for differential expression analysis of digital gene expression data. Bioinformatics **26**, 139–140 (2010)
26. Boyle, E.I., et al.: GO: termFinder–open source software for accessing Gene Ontology information and finding significantly enriched Gene ontology terms associated with a list of genes. Bioinformatics **20**, 3710–3715 (2004)
27. Azimi, S., et al.: Presence of exoY, exoS, exoU and exoT genes, antibiotic resistance and biofilm production among Pseudomonas aeruginosa isolates in Northwest Iran. GMS Hyg Infect Control **11**, Doc04 (2016)
28. Hoang, T.M., et al.: The heme-responsive PrrH sRNA regulates Pseudomonas aeruginosa pyochelin gene expression. mSphere **8**, e0039223 (2023)
29. Jia, X., et al.: Structural basis of sRNA RsmZ regulation of pseudomonas aeruginosa virulence. Cell Res. **33**, 328–330 (2023)
30. Shao, X., et al.: Novel therapeutic strategies for treating pseudomonas aeruginosa infection. Expert Opin. Drug Discov. **15**, 1403–1423 (2020)
31. Khaledi, A., et al.: Transcriptome profiling of antimicrobial resistance in pseudomonas aeruginosa. Antimicrob. Agents Chemother. **60**, 4722–4733 (2016)
32. Jurado-Martin, I., Sainz-Mejias, M., McClean, S.: Pseudomonas aeruginosa: an audacious pathogen with an adaptable arsenal of virulence factors. Int. J. Mol. Sci. **22**, 3128 (2021)

33. Oliveira, V.C., et al.: Expression of virulence factors by pseudomonas aeruginosa biofilm after bacteriophage infection. Microb. Pathog. **154**, 104834 (2021)
34. Esani, S., Chen, T., Leung, K.P., Van Laar, T.A.: Transcriptome sequence of antibiotic-treated pseudomonas aeruginosa. Microbiol Resour Announc **8**, e01367-e1418 (2019)

Feature Engineering on LMS Data to Optimize Student Performance Prediction

Keith Hubbard[✉] and Sheilla Amponsah

Stephen F. Austin State University, Nacogdoches, TX, USA
hubbardke@sfasu.edu

Abstract. Nearly every educational institution uses a learning management system (LMS), often producing terabytes of data generated by thousands of people. We examine LMS grade and login data from a regional comprehensive university, specifically documenting key considerations for engineering features from these data when trying to predict student performance. We specifically document changes to LMS data patterns since Covid-19, which are critical for data scientists to account for when using historic data. We compare numerous engineered features and approaches to utilizing those features for machine learning. We finish with a summary of the implications of including these features into more comprehensive student performance models.

Keywords: College Student Success · Learning Management System Data · Predictive Modeling · LMS Feature Engineering

1 Introduction

College student success and attrition, along with their connection to student engagement have been a frequent topic of research analysis (Kuh et al. 2008; Kuh et al. 2010; Hutt et al. 2018; Macnamara and Burgoyne, 2023). Employing data science on these topics is logical since the volume of data is so large. There are classroom-level learning analytics approaches focusing on a single context as well as student and faculty experiences with the data analysis (eg. Arnold and Pistilli, 2012; Dietz-Uhler and Hurn, 2013). There has been work on unpacking the data within learning management systems (LMS) to support student success (eg. Broadbent, 2016). However, Marques and colleagues' analysis concluded that, in their context, LMS data alone was insufficient to meaningfully predict student success (2017). Early alert systems are now used extensively across higher education as summarized in Velasco's metanalysis of the systems. Her research concludes that results vary with timing, staff training, and a variety of other factors, implying that the details of an institution's data utilization and interventions matter far more than just 'having an alert system' (2020).

In another approach to leveraging student data, Wong attributed a 4% increase in first-year retention at least partially to utilizing students' incoming data, orientation attendance, survey responses, and midsemester grades in a dashboard available to staff (2021). Other aspects of college students' data footprint have also been studied. College students' meal plan usage has been used to model student social networks (Bowman et al. 2019). Another study highlights the potential for meal usage data to support interventions for student success *in theory*, but were not able to produce significant predictors (Samuel and Scott, 2014). DeCarbo credits predictive analytics with improving course outcomes at one institution (without detail on methodology) but also concluded only a fifth of higher education institutions are identifying students who are at risk academically on any widespread level (2022). Pistilli's analysis shows how hard it is to utilize most predictive data on a timeline that maximizes its utility (2017). It seems that the underlying theme is that although extensive data is being gathered on students, optimizing its use requires attention to the specifics of that data.

Crossley et al. examined MOOC data for 320 students in a 2013 course called Big Data in Education. They analyzed click-stream data and natural language processing. They were able to predict with 78% accuracy which students would successfully complete the course.

Bird and colleagues examined a state-wide community college system attempting to create models to predict whether a particular student would pass a particular course. They compared using administrative data with using LMS data, and also a combination of the two. They also considered new students separately from returning students, finding that the LMS data was most helpful in improving predictions for new students, but minimally helpful for returning students (in press). This was a large study incorporating 226,784 students. Data collection ended in Spring 2021 (when only 75% of their courses contained any LMS activity), so the study is recent, but our research points to changes in LMS usage even since then.

The rapid growth of LMS's, and online learning in general, is well documented. An estimated 98% of universities moved online in April 2020 due to the covid-19 pandemic (Sadler, 2021). Although many institutions largely returned to face-to-face education, particularly in 2022, Bouchrika documents how LMS usage is again growing rapidly at universities (as high as 17% annually in Asia) (2025).

This article has three focuses:

- Examine how LMS Login and Grade data have changed over recent years at our institution, which has implications for training models on historic data;
- Examine how feature engineering affects the predictive power of LMS Login and Grade data – arguably the two most important features within LMS's;
- Examine how LMS data improves student performance models utilizing general student data.

2 Method

The research group from which this work emerges has grade data on all undergraduate students at a midsized regional comprehensive university dating back to 2010. We also have comprehensive LMS data dating back to 2016, however the nature of this data changes dramatically over time. We begin our methodology by explaining the rationale for the time periods on which our study focused. We then move to the rationale for the LMS features we will focus on in this work.

2.1 Timeframe Selection

Recall that Spring 2020 was the term when Covid-19 led to the reformatting of college courses worldwide. Our institution was already regularly utilizing an LMS system prior to 2020, so the change might not be as drastic as other institutions, but Table 1 outlines the change in LMS "Logins per Student" at the time of the Covid-19 outbreak. We use this table as a rationale to limit our attention to Fall 2020 through Fall 2024 data, since on some level the logins per student appear to stabilize over that period.

Table 1. Students enrolled full-time (12 + hours) and LMS logins of those students

	Total Students	Total Logins	Logins per Student
Spring 2019	8,318	1,668,586	201
Fall 2019	9,140	1,916,127	210
Spring 2020	8,092	1,874,798	232
Fall 2020	8,809	2,252,395	256
Spring 2021	7,669	1,923,893	251
Fall 2021	8,236	1,932,580	235
Spring 2022	7,030	1,726,979	246
Fall 2022	7,764	1,655,993	213
Spring 2023	6,675	1,348,794	202
Fall 2023	7,735	1,861,928	241
Spring 2024	6,686	1,815,653	272
Fall 2024	8,101	2,171,345	268
Total	**94,255**	**22,149,071**	**235**

The table also restricts attention to "long-semesters," meaning no summer terms. We chose to restrict to these terms since course structures differ dramatically at our university over the summer (with much smaller numbers of students).

We first examined registered logins for the Brightspace Desire2Learn LMS over nine long semesters from Fall 2020 to Fall 2024. There were a total of 19,977,140 logins by 27,875 unique users over the period of consideration. We associated this data with information about the users. We then disaggregated these logins by term, by major, by gap between login, and by other measures attempting to understand which aspects of logins yielded the best predictive power for students' semester grade point average (GPA), their overall GPA, and their retention at the university.

We also examined all grades assigned within the LMS over the same period. There were a total of 10,129,981 grades assigned to total of 29,233 distinct students. Again, grades were associated with underlying student data, examined and disaggregated. Ultimately, we attempted to utilize these grades to optimize prediction of student GPA and retention at the university.

In total we considered 85,848 undergraduate student-semester outcomes from the nine semesters of interest. For each of these student-semesters, it would be useful to have predictions as to how the student performed (GPA) and whether the student was retained. A total of 23,471 distinct courses are represented.

2.2 LMS Feature Selection

The most dominant features in predicting student performance are prior student GPA and midsemester grades (which are provided by faculty to indicate what grades are likely to be at the end of the term). These features are generally widely available to university personnel and fairly simple to interpret.

In contrast, LMS data is *not* widely available to university personnel and not well understood in the aggregate. Specifically, at our university neither a student's academic advisor nor any of their individual professors has access to *all* of their LMS grades, discussion posts, logins, or any other features. By design, Learning *Management* Systems are supposed to track or manage the learning / performance a student does, so there is every reason to believe this data should be profoundly predictive of a student's academic performance. Our team examined 72 features from the LMS and settled on around 10 features that showed particular promise. These features were cleaned, missing data was analyzed, and the feasibility of resultant engineered features was analyzed. We delay detailed descriptions of how features were engineered until the Analysis section of this paper, but examine feature importance, first in terms of correlation to three outcomes of interest: Students' *semester GPA*, Students' *overall GPA* at the end of semester, and Students' *discontinuance* (or whether they left the university at the end of the semester without a degree). Table 2 presents the top 10 features correlated to each of the three outcomes of interest.

Table 2. Top 10 features, ranked by correlation to 3 key outcomes.

Semester GPA	Overall GPA	Discontinuance
Midterm grades	Beginning GPA	High school
LMS Grades	Midterm grades	**LMS Periodic Logins**
Beginning GPA	**LMS Grades**	Student type code
Alerts	High school	**LMS Grades**
High school	High school percentile	Major
LMS Periodic Logins	Alerts	Total hours attempted
High school percentile	Major	TSI score
Major	LMS content completion	Midterm grades
LMS content completion	**LMS Periodic Logins**	Beginning GPA
LMS content visits	LMS content topics	Admit code

Observe that "LMS grades" appears as the second, third, and fourth most significant predictors in these lists, while LMS Periodic Logins appears as the sixth, ninth, and second most significant logins. Other LMS features appear, but never ranking higher than seventh.

We also employed Recursive Feature Elimination (RFE) to a random forest model for each of the three outcomes of interest. We used a random forest model initially because nearly all models which have historically performed best on this type of data are aggregations of decision trees. Table 3 summarized the top five features on the three outcomes of interest. Observe that the only two LMS features to appear in Table 3 after RFE are LMS Grades and LMS Periodic Logins. With these justifications, we move onto our analysis of features and feature engineering pertaining the LMS grades and logins.

Table 3. Top 5 features, ranked by RFE on the 3 key outcomes

Semester GPA	Overall GPA	Discontinuance
Midterm grades	Beginning GPA	High school
LMS Grades	Midterm grades	**LMS Grades**
Beginning GPA	**LMS Grades**	Total hours attempted
High school	High school	Midterm grades
LMS Periodic Logins	Total hours complete	Beginning GPA

3 Analysis

We consider LMS login data, then LMS grade data. Finally, we summarize how these data combine with other student data for more comprehensive prediction models.

3.1 LMS Login Data

Calculating a correlation coefficient between the number of student logins during a term and their semester GPA yields 0.22. However, many logins are quite close together. By restricting our consideration to logins at least a minimum distance apart, we are able to improve the relationship to semester GPA as Fig. 1 illustrates. In fact, the highest correlation coefficient is 0.38 when removing all logins less than 11 h apart.

We also examined ignoring logins that were spaced beyond a given maximum. This produced marginal correlation gains but did not combine effectively with the substantial gains seen by imposing minimum login gaps. Additionally, we examined counting 24-h or 12-h periods with a login for each student. Neither performed as well as the minimum login gaps illustrated in Fig. 1.

As seen in Fig. 2, the predictive power of logins varies dramatically between different major types with the highest being 0.55 correlation for Psychology majors (peaking at 11 h gaps) and the lowest being 0.20 correlation for Dual Credit students (peaking at 10 h gaps). Note that all peak correlations occur between 10 and 11 h, indicating that this feature is robust across subpopulations.

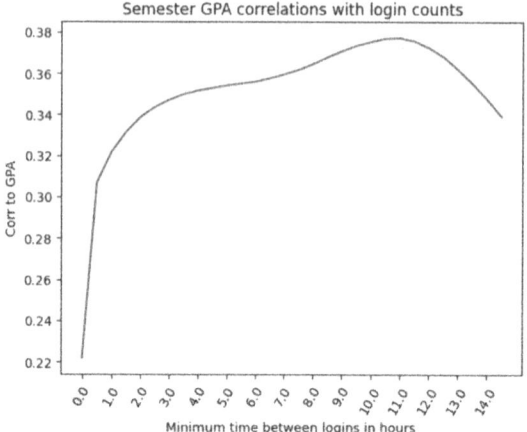

Fig. 1. Correlation between semester GPA and login counts, only counting logins with the specified minimum login gaps

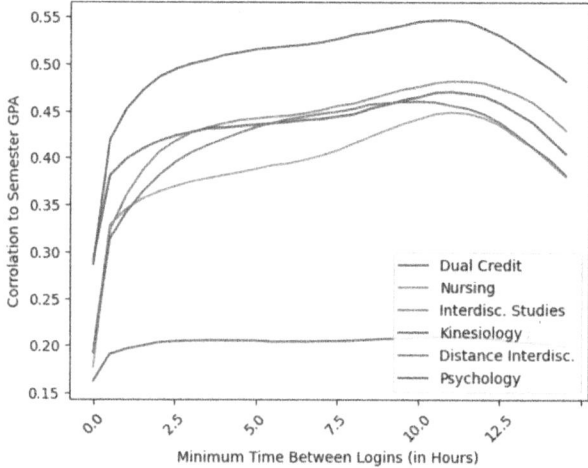

Fig. 2. Correlations between semester GPA and login counts, with specified minimum login spacings, disaggregated by the top 6 majors

We also examine the correlations of logins with credit hours attempted and credit hours completed as illustrated on the left and the right of Fig. 3, respectively. Students who attempted 15 or more credit hours had the highest correlation at 0.49 (peaking at 10.5 h gaps) and the lowest was 0.22 for students with at most 3 credit hours attempted (peaking at 11 h gaps). For credit hours completed, students having completed between 31 to 60 credit hours (peaking at 10.5 h gaps) had the highest correlation between periodic logins and semester GPA at 0.5. Tied for lowest correlation, 0.3, were students with at most 30 total credit or more than 90 total credit hours. Note that all peak correlations occur between 10.5 and 11.5 h, indicating that this feature is again robust across subpopulations.

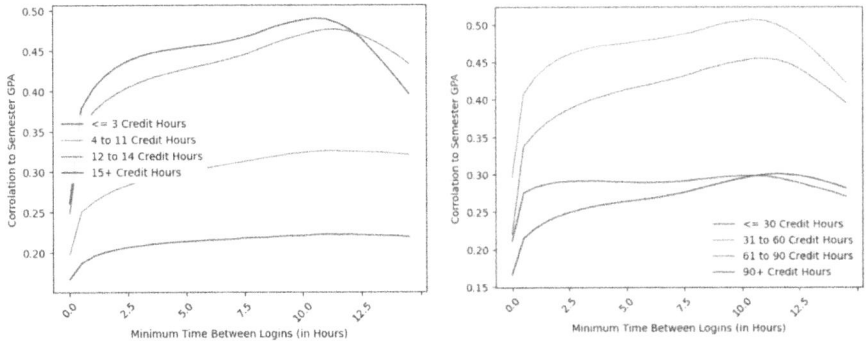

Fig. 3. Correlations between semester GPA and login counts, with specified minimum login spacings, disaggregated by the credit hours attempted on the left and credit hours completed on the right

We focused our efforts on interpreting the utility of periodic login counts with a minimum 11-h gaps between logins. Deploying flaml to test a variety of regressors to use this single variable to predict students' semester GPA, XGBoost regressor performed best. It produced a prediction on the test set (which was 20% throughout) an MSE of 0.818; 42% of the test set GPAs were predicted within 0.5 grade point; and 77% of the test set were predicted within 1 grade point.

Since predictive power is more important early in the semester for students, we also restricted to periodic logins in the first 4 weeks, first 8 weeks, and first 12 weeks (not counting Spring Break), training models on each. We also built one model using all four timeframes to allow for the possibility that a change in the number of logins throughout the semester was a significant feature. As seen in Table 4, the models were not wildly disparate. Notice that all of the models are decisions tree aggregators and four-of-five employ gradient boosting. Models were compared on 1- R2. For performance comparison, on the "All Combined" modeling LMGM did best at 0.7937, then RF at 0.7955, then XGBoost at 0.8044, then ExtraTrees at 0.982, then XGB_LimitDepth at 0.9068, then SGD at 0.9585.

Table 4. The performance of periodic login data from different portions of the semester in predicting semester GPA

	First 4 Weeks	First 8 Weeks	First 12 Weeks	Full Semester	All Combined
Periodic logins	1,849,465	3,475,366	5,173,306	6,705,135	n/a
MSE	0.899	0.882	0.874	0.818	0.760
within 0.5 point	40%	41%	41%	42%	43%
within 1 point	76%	77%	77%	77%	79%
Best model	LGBM	LGBM	ExtraTrees	XGB	LGBM

Login data was less effective in predicting discontinuance. Again utilizing flaml, we examined classification models over the same time frames as documented in Table 5. These is some improved prediction power with more weeks of data, but none of the numbers are impressive. Again, models performed moderately similarly (comparison using area 1 – "area under ROC"). On the "All Combined" model, XGBoost performed best (0.2498), then RF (0.2510), then LGBM (0.2523), then XGB_LimitDepth (0.2565), ExtraTrees (0.2566), then SGD (0.3446).

Table 5. The performance of periodic login data from different portions of the semester in predicting discontinuance

	First 4 Weeks	First 8 Weeks	First 12 Weeks	Full Semester	All Combined
Accuracy	0.89	0.89	0.89	0.89	0.89
Area under ROC	0.67	0.70	0.73	0.74	0.76
Best model	ExtraTrees	ExtraTrees	RF	LGBM	XGB

Note that in the context of identifying and connecting with students who are discontinuing, the overall ROC curve is less critical than the endpoints. Judged through a student retention lens, for expensive student interventions one requires high precision and for inexpensive / bulk interventions really only a high recall is required

3.2 LMS Grade Data

We now turn our attention to grade data within the LMS. Instructors have the ability to add every grade, numeric or not, into the LMS to allow students to monitor their progress. Most universities do not have a requirement that all, or even any, course grades be entered into an LMS. The university under consideration is no different.

Table 6 summarizes number of students, courses, and LMS grades across terms. Observe that student count is down slightly over times, and grade volume is up. Combined, we see that the grades-per-student average over the first two terms in the dataset is 90, while in the last two terms an average of 136 grades-per-student are recorded, a 51% increase. This increase resulted almost entirely from grades that were updated before the end of the semester. The increase in usage likely came first from the push to LMS communication in the immediate aftermath of Covid-19, but then moved earlier in the term as faculty more robustly incorporated LMS grading into their courses rather than just posting grades toward the end of the term.

We retained all copies of updated grades, since the timing of these updates (and the original grades) comes into play with time sensitive grade models. If only unique grades were counted, there would be 7,667,032 in total, an 89.8 grades-per-student average to start and a 90.4 average at the end.

LMS grades are assigned throughout the semester, with earlier grades providing more utility for improving student outcomes while there still might be time to support struggling students. Within each timeframe in the right three columns of Table 6, we eliminated duplicate grades for a given grade item, keeping the last awarded grade. Figure 4 depicts the volume of grades-per-student entered within four-week timeframes. The proportion of grades assigned early in the semester has increased over time, supporting our earlier hypothesis about change in faculty usage. Specifically, the first two terms average 12.2 grades-per-student in the first half of the term while the last terms averaged 20.8, a 36% increase.

Table 6. Semester totals for student, courses, and LMS grades over nine semesters

	Students	Courses	LMS Grades	Unique Grades by Period:		
				First 4 Weeks	Next 4 Weeks	Rest of Term
Fall 2020	10,898	2,742	995,813	167,679	244,042	578,530
Spring 2021	9,784	2,880	871,157	106,057	154,872	607,336
Fall 2021	10,216	2,626	884,789	158,707	220,272	503,869
Spring 2022	8,971	2,683	1,164,765	177,163	188,921	448,867
Fall 2022	9,742	2,562	1,262,091	182,727	208,561	447,711
Spring 2023	8,614	2,481	1,159,832	171,877	168,436	418,519
Fall 2023	9,457	2,505	1,314,557	198,004	220,073	453,159
Spring 2024	8,374	2,429	1,185,588	178,221	165,364	439,126
Fall 2024	9,792	2,563	1,295,260	199,870	221,158	438,091
Total	85,848	23,471	10,133,852	1,540,305	1,791,699	4,336,397

Grades were assigned in a variety of ways. Our LMS allows for a "Points Numerator", a "Points Denominator", and / or a "Grade Value" for each grade. "Points Numerator" had 86,940 missing values, "Points Denominator" had 62,299 missing values, and "Grade Value" had 3,871 missing values and 1,364,804 non-float values. Hence, data analysis considered each of these as a predictor.

Since the numerator and denominator version of each grade has most numeric values (and an inherent weighting of grades given by denominator) we examined these by summing numerators, summing denominators, and dividing for each student (without regard to course). This minimized missing data since a student had a grade average if *any* of their courses had grades. For the remaining missing grades, we used median imputation, as we will throughout this analysis. We then employed machine learning to predict student semester GPA base *only* on LMS grades. We utilized flaml, which attempted to fit six models. ExtraTrees achieved the best prediction of semester GPA with an MSE of 0.455, 62% of the test set GPAs predicted within 0.5 grade point, and 85% predicted within 1 grade point.

On the other hand, converting the top 8 non-float Grade Values to floats (such as 95 for "A"), averaging all grades for each student, then executing machine learning also yielded best results using LGBM with an MSE of 0.385, 67% of the test set GPAs within 0.5 grade point, and 90% within 1 grade point. This is despite the fact that not every course has *any* LMS grades entered. So far, we are including even late semester grades which have limited utility in helping students change course. (This is the rightmost column of Table 7.) For comparison, the performance, as measured by $1-R^2$, was LGBM (0.3995), then RF (0.4001), then XGB_LimitDepth (0.4025), then XGBoost (0.4030), then ExtraTrees (0.4044).

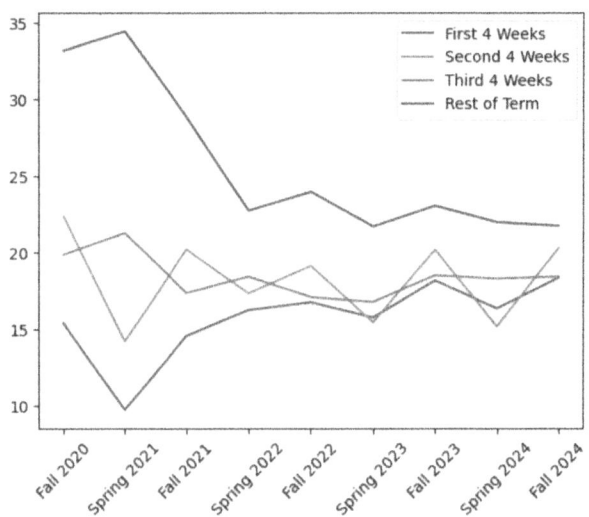

Fig. 4. Unique grades per student across terms, disaggregated by the four-week period in which they were posted

Next, we recalculated both numerator-over-denominator and grade-average, first averaging all graded courses for each individual, then by weighting the average of those outcomes by the credit hours of each course. For the weighted numerator-over-denominator, LGBM performed best with an MSE of 0.406, 66% of the test set accurate to within 0.5 grade point, and 89% accurate to within 1 grade point. For the weighted grade-average, the best performance had an MSE of 0.420, 65% of the test set accurate to within 0.5 grade point, and 89% accurate to within 1 grade point.

Finally, considering the fact that at times optional assignments have points denominators (which would previously have effectively been assumed to be 0 if not completed), we calculated a median number of points for each course. We then divided each student-course numerator by the course median. We averaged across courses for each student, then calculated a weighted average across courses for each student. For the unweighted numerator-over-median, the best performing model, LGBM, had an MSE of 0.428, 61% of the test set accurate to within 0.5 grade point, and 89% accurate to within 1 grade point. For the weighted grade-average, now XGBoost performed best with an MSE of 0.398, 66% of the test set accurate to within 0.5 grade point, and 89% accurate to within 1 grade point.

Our perhaps unintuitive conclusion is that weighting grades by the denominator (the points declared possible by the instructor) and weighting grades in classes by the credit hours in that class both actually *decrease* the predictive power of grade data in this dataset. We restrict our attention to only the best feature.

We proceed to examine partial term data, since educational institutions want to know which students are predicted not be successful as early in the term as possible. We will restrict our attention to unweighted grade value data (with the most frequent string values converted to numeric, ex. A becomes 95). As seen in Table 7, grade data within the first 4 weeks allows for 80% accurate GPA prediction within 1 grade point, raising to 90% accuracy by the end of the term.

Table 7. The performance of grade data from different portions of the semester in predicting semester GPA

	First 4 Weeks	First 8 Weeks	First 12 Weeks	Full Semester
# of grades	1,832,999	4,275,656	6,797,565	10,133,852
MSE	0.762	0.598	0.491	0.385
within 0.5 point	48%	57%	62%	67%
within 1 point	80%	84%	87%	90%
Best model	LGBM	LGBM	LMGM	LMGM

These early semester predictive findings are surprisingly good and seem to contradict Marques and colleagues' conclusion that LMS data alone was inadequate to meaningfully predict student success (2017). It is probable that LMS usage on university campuses has changed dramatically since 2017. More comparative work would be necessary to determine whether that was the primary difference in findings. Of critical importance, however, is the fact that 4 weeks of LMS grade data were sufficient to estimate 80% of students' semester GPA within one grade point.

Discontinuing college is more challenging to predict. Using the 4-week, 8-weeks, 12-week, and full grade data we also trained models with discontinuance as the outcome. Figure 5 outlines the different ROC curves yielded from the 4 different data levels. Note that the area under curve (AUC) increases with each additional influx of data, but the increase is not dramatic. Best performance, as measured by 1-AUC, was achieved by XGBoost and XGB_LimitDepth (both 0.2880), followed by LGBM (0.2882), then ExtraTrees (0.2888), then RF (0.2896), then SGD (0.4968).

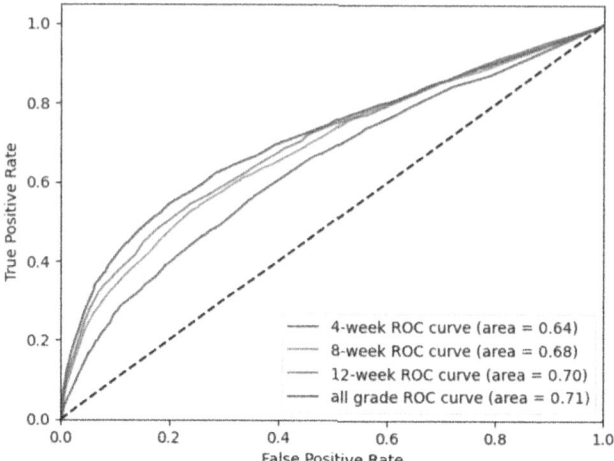

Fig. 5. ROC Curves for predicting student discontinuance based on varying subsets of LMS grade data

Integrating LMS Data with Traditional Student Predictors

By far the most common approach with LMS data is to combine it with other data sources to predict student performance. For instance, Bird and colleagues incorporated 250 course-specific predictors (in press).

In this analysis, we focus on midsemester analysis and use 20 factors (which produces a 1–2% improvement over 10 factors). The semester GPA prediction that performed best was the XGB_LimitDepth regressor as seen in Table 8. It achieved an MSE of 0.315, an R^2 of 0.6823, predicted 73% of the test set within 0.5 grade point of their GPA, and predicted 92% within 1 point of their GPA. Performance of the next best model, XGBoost, was about 1.5% worse using $1 - R^2$. Optimal configurations for each model can be found in the Appendix.

Table 8. Performance of various models predicting the 3 key outcomes using 20 features

	Semester GPA $(1-R^2)$	Overall GPA $(1-R^2)$	Discontinuance $(1-AUC)$
LGBM	0.3231	0.1482	0.1433
Random Forest	0.3321	0.1641	0.1505
XGBoost	0.3226	0.1477	0.1543
ExtraTrees	0.3299	0.1531	0.1499
XGB_LimitDepth	**0.3177**	**0.1451**	**0.1428**
SGD	0.7532	0.7933	0.3303
LRL 1			0.2455

As seen in Table 8, predicting overall GPA was best modeled using the XGB_LimitDepth regressor, which achieved an MSE of 0.095, an R^2 of 0.8549, predicted 93% of the test set within 0.5 grade point of their GPA, and predicted 98% within 1 point of their GPA. The performance of the next best model, XGBoost, was 1.8% worse.

Finally, the best model found for discontinuance was XGB_LimitDepth classifier. It achieved accuracy of 0.91 and had an area under ROC of 0.8572. The second-best model, LGBM, was only 0.35% behind as seen in Table 8.

We calculated SHAP values on all three models for the approximately 17,000 values in the test set. As seen in Fig. 6, the mean influence of LMS logins and grades on the test set was substantial – the 2nd and 4th most significant features for semester GPA, the 3th and 5th for overall GPA, and 2nd and 6th for discontinuance.

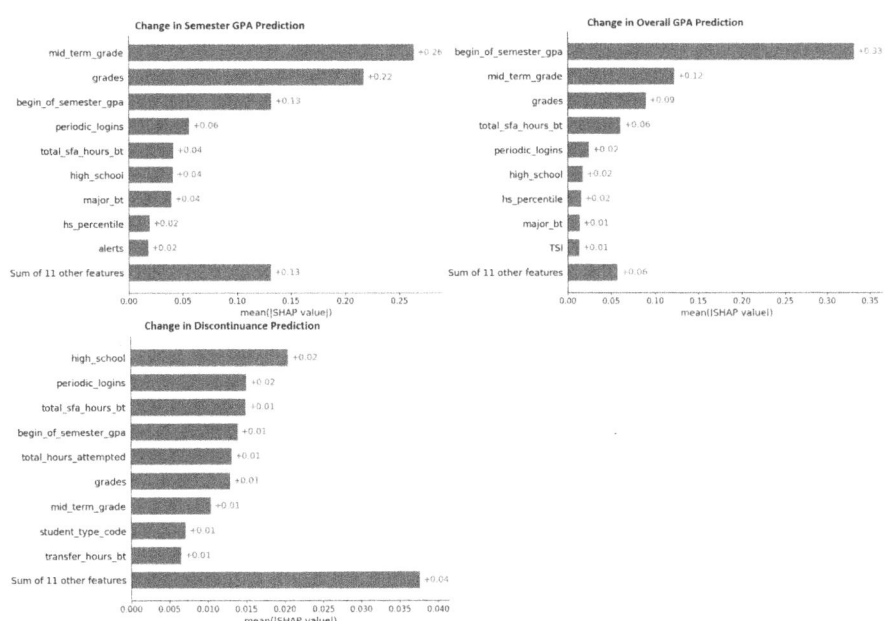

Fig. 6. Mean absolute SHAP values of features for 3 key outcomes

4 Applications

These predictive models are of great interest to university stakeholders on our campus. Since these models can be informed by midsemester LMS data, the primary model in practice, a semester GPA predictor, is updated every Monday and provided to academic advisors that serve approximately 60% of undergraduate students. Another cabinet-level administrator is also interested in leveraging these predictive models to inform Student Affairs interventions. They hope to make use of the overall GPA model, but may also use the discontinuance model.

Currently, raw counts of LMS logins, tutor visits, and other dynamic features are provided to advisors to help them interpret the model's predictions. However, the authors are exploring the deployment of SHAP values to assist student support staff in interpreting model predictions for specific students.

Finally, there appears to be tremendous interest in models for specific populations for which (a) predictions are particularly accurate such as full-time students, Psychology students, or sophomores and juniors (as illustrated in Figs. 2 and 3); or (b) specific interventions might be particularly useful (such as career coaching for students not on track to be accepted into the nursing program). Work on specialized models for subpopulations is ongoing.

5 Conclusion

LMS data provides vital insight into the academic well-being of students midsemester. The quality of the analysis of these data can dramatically increase their predictive power, as seen especially with logins versus periodic logins. Further, LMS usage has changed in our context, even since 2021. Accounting for differences between legacy data will be particularly relevant for LMS data. The intention of this article is not to *tell* other institutions which format of their data will be most predictive, but to *encourage* other institutions to carefully examine their unique data characteristics and to provide ideas about key aspects to analyze.

Different engineered features appear to have different strengths, but in the present dataset it appears that attempts to weight LMS grades are counterproductive. Grades across classes seem to do a good job of predicting overall GPA performance, even if some classes are not represented by grades in the LMS. LMS logins, however, appear to do a better job of predicting discontinuance from college.

LMS data is a vital tool in improving overall predictive models for student success, but may be among the most sensitive features because of their volume, the variety of who entered them, and their multifaceted interpretations. In the future, it is likely that learning management systems will be used even more heavily in education. The data science community would be wise to continue to study and share insights on this important topic.

Acknowledgments. This study was made possible in part by a faculty development leave award from Stephen F. Austin State University.

Disclosure of Interests. The authors have no competing interests to declare that are relevant to the content of this article.

Appendix: Best Configurations of Estimators Considered

Appendix: Best Configurations of Estimators Considered

Prediction 1: semester_gpa

Features: mid_term_grade, grades, begin_of_semester_gpa, alerts, high_school, periodic_logins, hs_percentile, major_bt, content_completion_rate, content_topic_visits, TSI, college_bt, new_SAT, admit_code , total_time, transfer_hours_bt, total_sfa_hours_bt, student_type_code, meal_plan, hall_name

LGBM best configuration: {'n_estimators': 1173, 'num_leaves': 10, 'min_child_samples': 6, 'learning_rate': 0.054987380085078814, 'log_max_bin': 9, 'colsample_bytree': 0.885652480919423, 'reg_alpha': 0.009984034615570498, 'reg_lambda': 0.2388700003862822}

RF best configuration: {'n_estimators': 63, 'max_features': 0.6739758718573503, 'max_leaves': 1057}

XGBoost best configuration: {'n_estimators': 442, 'max_leaves': 360, 'min_child_weight': 20.751061790819115, 'learning_rate': 0.03399093100192437, 'subsample': 0.8809985595678514, 'colsample_bylevel': 0.8490633393816673, 'colsample_bytree': 0.7514653053266084, 'reg_alpha': 0.006254798766353442, 'reg_lambda': 0.28534114950225703}

ExtraTrees best configuration: {'n_estimators': 71, 'max_features': 0.8316456275359224, 'max_leaves': 4760}

XGB_LimitDepth best configuration: {'n_estimators': 2199, 'max_depth': 10, 'min_child_weight':6.137984684618797, 'learning_rate': 0.007759642790692926, 'subsample': 1.0, 'colsample_bylevel': 0.7316290044540736, 'colsample_bytree': 0.8903447372116292, 'reg_alpha': 0.0010515900751535718, 'reg_lambda': 4.050395673709929}

SGD best configuration: {'penalty': 'l2', 'alpha': 2.073827860066681e-05, 'l1_ratio': 0.9999999999999999, 'epsilon': 0.04721809031174382, 'learning_rate': 'optimal', 'eta0': 0.013646120811719356, 'power_t': 0.27563902198714846, 'average': False, 'loss': 'epsilon_insensitive'}

Prediction 2: end_of_term_gpa

Features: begin_of_semester_gpa, mid_term_grade, grades, hs_percentile, high_school, alerts, major_bt, periodic_logins, content_completion_rate, content_topic_visits, TSI, new_SAT, college_bt, admit_code, student_type_code, transfer_hours_bt , total_time, total_sfa_hours_bt, meal_plan, hall_name

LGBM best configuration: {'n_estimators': 557, 'num_leaves': 17, 'min_child_samples': 15, 'learning_rate': 0.042242947825983146, 'log_max_bin': 9, 'colsample_bytree': 0.5623352336965426, 'reg_alpha': 0.0009765625, 'reg_lambda': 0.07755818552475893}

RF best configuration: {'n_estimators': 162, 'max_features': 0.584414084625783, 'max_leaves': 2263}

XGBoost best configuration: {'n_estimators': 950, 'max_leaves': 60, 'min_child_weight': 7.233785289465286, 'learning_rate': 0.015126332186899651, 'subsample': 1.0, 'colsample_bylevel': 0.3685988387608087, 'colsample_bytree': 0.8129075176371072, 'reg_alpha': 0.0019706937086253853, 'reg_lambda': 0.1148611342151243}

ExtraTrees best configuration: {'n_estimators': 250, 'max_features': 0.7088416535926986, 'max_leaves': 12114}

XGB_LimitDepth best configuration: {'n_estimators': 228, 'max_depth': 13, 'min_child_weight': 40.26812114743653, 'learning_rate': 0.06219377762209235, 'subsample': 0.8365749881305449, 'colsample_bylevel': 0.8488288054460789, 'colsample_bytree': 0.785498557164527, 'reg_alpha': 0.09749190094363033, 'reg_lambda': 47.50558410170101}

SGD best configuration: {'penalty': 'l2', 'alpha': 3.57153377370962e-05, 'l1_ratio': 0.9999999999999999, 'epsilon': 0.006407201316208552, 'learning_rate': 'optimal', 'eta0': 0.008615774063807258, 'power_t': 0.3052063818916277, 'average': False, 'loss': 'epsilon_insensitive'}

Prediction 3: discontinued

Features: high_school, periodic_logins, student_type_code, grades, major_bt, total_hours_attempted, TSI, mid_term_grade, begin_of_semester_gpa, admit_code, total_sfa_hours_bt, college_bt, content_topic_visits, alerts, content_completion_rate, transfer_hours_bt, hs_percentile, total_meals, first_gen_status, meal_plan

LGBM best configuration: {'n_estimators': 2541, 'num_leaves': 1667, 'min_child_samples': 29, 'learning_rate': 0.0016660662914022304, 'log_max_bin': 8, 'colsample_bytree': 0.5157078343718623, 'reg_alpha': 0.045792841240713165, 'reg_lambda': 0.0012362651138125363}

RF best configuration: {'n_estimators': 1000, 'max_features': 0.1779692423238241, 'max_leaves': 7499, 'criterion': 'gini'}

XGBoost best configuration: {'n_estimators': 13499, 'max_leaves': 60, 'min_child_weight': 0.008494221584011285, 'learning_rate': 0.006955765856675575, 'subsample': 0.5965241023754743, 'colsample_bylevel': 0.590641168068946, 'colsample_bytree': 1.0, 'reg_alpha': 0.2522240954379289, 'reg_lambda': 5.351809144038808}

ExtraTrees best configuration: {'n_estimators': 2047, 'max_features': 0.46132798093546956, 'max_leaves': 12856, 'criterion': 'gini'}

XGB_LimitDepth best configuration: {'n_estimators': 877, 'max_depth': 11, 'min_child_weight': 0.6205465771093738, 'learning_rate': 0.013622118381700795, 'subsample': 0.566692814245426, 'colsample_bylevel': 0.8865741642101924, 'colsample_bytree': 1.0, 'reg_alpha': 0.01386336444764391, 'reg_lambda': 3.113947886074155}

SGD best configuration: {'penalty': 'l2', 'alpha': 0.0001, 'l1_ratio': 0.1500000000000002, 'epsilon': 0.1, 'learning_rate': 'invscaling', 'eta0': 0.010000000000000005,

'power_t': 0.5, 'average': False, 'loss': 'modified_huber'}

LRL_1 best configuration: {'C': 1.0}

References

Arnold, K.E., Pistilli. M.D.: Course signals at purdue using learning analytics to increase student success. Proceedings of the 2nd International Conference on Learning Analytics and Knowledge (2012)

Barley, Z., Brigham, N.: Preparing teachers to teach in rural school. Issues and Answers, Institute of Education Sciences, REL 2008 – No. 45 (2008)

Bird, K.A., Castelman, B.L., Song, Y., Yu, R.: Is Big Data Better? LMS Data and Predictive Analytic Performance in Postsecondary Education. EdWorkingPapers (in press) https://edworkingpapers.com/ai22-647

Bouchrika, I.: 51 LMS Statistics: 2025 Data, Trends & Predictions. Research.com (2015). https://research.com/education/lms-statistics

Bowman, N., Jarratt, L., Polgreen, L.A., Kruckeberg, T., Segre, A.M.: Early identification of students' social networks: predicting college retention and graduation via campus dining. J. Coll. Stud. Dev. **60**(5), 617–622 (2019)

Broadbent, J.: Academic success is about self-efficacy rather than frequency of use of the learning management system. Australasian Journal of Educational Technol. **32**(4) (2016)

Crossley, S., Paquette, L., Dascalu, M., McNamara, D., Baker, R.: Combining click-stream data with NLP tools to better understand MOOC completion. ACM Digital Library 6–14, (2016) https://doi.org/10.1145/2883851.2883931

DeCarbo, B.: How Data Is Changing the College Experience. Wall Street Journal, August 19 (2022)

Dietz-Uhler, B., Hurn, J.E.: Using learning analytics to predict (and improve) student success: a faculty perspective. Journal of Interactive Online Learning **12**, 17–26, Spring (2013)

Donachie, P.: Rural schools are lacking necessary resources, report finds. K-12 Dive, June 14 (2017) https://www.k12dive.com/news/rural-schools-are-lacking-necessary-resources-report-finds/444933/

Goodpaster, K.P.S., Adedokun, O.A., Weaver, F.C.: Teachers' perceptions of rural STEM teaching: implications for rural teacher retention. The Rural Educator **33**(3), 9–22 (2015)

Hutt, S., Gardener, M., Kamentz, D., Duckworth, A.L., D'Mello. S.K.: Prospectively predicting 4-year college graduation from student applications. Proceedings of the 8th International Conference on Learning Analytics and Knowledge, Assoc. for Computing Machinery, pp. 280–289 (2018)

Kuh, G.D., Kinzie, J., Schuh, J.H., Whitt, E.J.: Student Success in College: Creating Conditions That Matter. San Francisco, CA (2010)

Kuh, G.D., Cruce, T.M., Shoup, R., Kinzie, J., Gonyea, R.M.: Unmasking the effects of student engagement on first-year college grades and persistence. J. Higher Educ. **79**(5), 540–563 (2008)

Macnamara, B.N., Burgoyne, A.P.: Do growth mindset interventions impact students' academic achievement? A systematic review and meta-analysis with recommendations for best practices. Psychol. Bull. **149**(3–4), 133–173 (2023)

Marques, B.P., Villate, J.E., Carvalho, C.V.: Analytics of student behaviour in a learning management system as a predictor of learning success. 12th Iberian Conference on Information Systems and Technologies (2017)

Monk, D.H.: Recruiting and retaining high-quality teachers in rural areas. Future Child. **17**(1), 155–174 (2007)

Sadler, M.: 49 LMS Statistics and Trends for a Post-COVID World. Trust Radius (2021) https://solutions.trustradius.com/vendor-blog/lms-statistics-trends/

Samuel, K.R., Scott, J.: Promoting hispanic student retention in two texas community colleges. Research in Higher Education Journal **25**, NP (2014)

Velasco, J.M.: Early Alert Programs: A Closer Look. Dissertation. Valdosta State University. https://www.proquest.com/docview/2419104942 (2020)

Wong, W.: Higher education turns to data analytics to bolster student success. Ed Tech Magazine, October 12 (2021)

Yu, R., Li, Q., Fischer, C., Doroudi, S., Xu D. : Towards accurate and fair prediction of college success: evaluating different sources of student data. In: Rafferty, A.N., Whitehill, J., Cavalli-Sforza, V., Romero, C. (eds.) Proceedings of the 13th International Conference on Educational Data Mining, pp. 292–301 (2020)

Socioeconomic Effects on Health and Well-Being Using U.S. County-Level Data

Lily Shaw, Erdogan Dogdu[✉], Roya Choupani, Steven Womack, and Minh Le

Department of Computer Science, Angelo State University, San Angelo, TX, USA
{mshaw7,edogdu,rchoupani,swomack5,hle2}@angelo.edu

Abstract. In this paper we analyze publicly available US county-level data on social demographics, behaviors, and health outcomes to investigate correlations between various health factors and outcomes. Specifically, we focus on exploring possible links between socioeconomic factors, race/ethnicity, and health outcomes with the provided US county data. Our study presents findings through correlational matrices and heatmaps, visually representing the relationships. We investigate correlations such as the impact of negative economic factors on health outcomes compared to areas with better economic conditions. Additionally, we examine the relationship between educational attainment and health outcomes, and how income inequality and childhood poverty affect health behaviors like smoking and alcohol intake. We also explore how unemployment and lower insurance rates may influence mental and physical health outcomes differently. Furthermore, we analyze whether areas with higher ratios of primary care physicians and mental health providers experience more teen births. We also investigate how changes in unemployment and educational attainment may lead to shifts in health outcomes. Finally, we assess how the physical environment of a county affects the overall health outcomes of its population.

Keywords: healthcare · data science · data analysis · predictive analytics

1 Introduction

Health outcomes are influenced by a multitude of factors, ranging from individual behaviors to broader societal structures. Among these determinants, socioeconomic status stands out as a crucial predictor, exerting a profound impact on an individual's access to healthcare, lifestyle choices, and overall well-being. Understanding the complex interplay between socioeconomic factors and health outcomes is paramount for addressing health disparities and promoting equitable access to healthcare resources.

In the United States, despite advancements in medical technology and healthcare delivery, significant disparities persist in health outcomes across different regions and demographic groups. While some areas boast high life expectancy and low rates of chronic diseases, others struggle with disproportionately high mortality rates and a prevalence of preventable health conditions. These disparities are reflective of underlying socioeconomic inequalities, including disparities in income, education, employment opportunities, and access to healthcare services.

Against this backdrop, our study seeks to delve into the nuances of socioeconomic disparities in health outcomes across the United States, employing a regionalized approach to explore variations in health outcomes and their correlates. By examining data from diverse geographical regions, we aim to uncover patterns, identify disparities, and elucidate the complex web of factors influencing health outcomes at the regional level.

Through comprehensive analysis and exploration, our research endeavors to contribute to the existing body of knowledge on health disparities and inform evidence-based strategies for addressing these inequities. By shedding light on the geographic variations in health outcomes and their underlying determinants, we aspire to provide valuable insights for policymakers, researchers, healthcare providers, and other stakeholders striving to promote health equity and improve the well-being of all Americans.

In this paper, we present our analysis of US County Health Data (2023) provided by the County Health Rankings & Roadmaps (CHR&R) program[1], a project by the University of Wisconsin Population Health Institute [1]. We discuss our methodology and criteria for input selection for both exploratory data analysis and predictive modeling. The detailed methodology for our analysis is provided in Section 3 of this paper. Most of our work and research for this paper were conducted using Python tools and the Pandas library, leveraging both Google Colab and Jupyter Notebooks.

The findings from our exploratory data analysis are presented and analyzed using standard Pearson correlations, along with examination of maximums/minimums and heatmaps to elucidate potential insights. Additionally, we will include other methods such as ranked correlations, albeit in a smaller capacity. For predictive modeling, our findings will primarily feature linear regressions. However, we also discuss and attempted other methods.

Section 2 presents related work in this research area. The methods we employed analyzing health data is presented in Sect. 3 and results are presented in Sect. 4. We conclude in Sect. 5 and point to future work.

2 Related Work

There have been many related works using the same county level health data, provided by County Health Rankings (CHR). Many papers seek to review the status of health and wellbeing in the U.S. and possible relationships between those outcomes and input factors. Ours is a new effort to identify relationships between socioeconomic factors and health outcomes in US-wide and regional levels.

Most papers have focused on analyzing the data and any correlations between factors, but also on the data collection itself. Due to the nature of the data being collected, it is frequently imperfect. Problems of underrepresentation in poorer areas of communities is frequently seen throughout the county health rankings. Many places do not collect certain racial or socioeconomic data. The reasoning for the lack of collection can be speculated on, yet the lack of data still exists. As such, most studies have chosen to primarily focus on those categories/factors that have a 90–95% completion rate. Arndt et al. investigates the reliability of country and regional health rankings in an earlier study

---

[1] https://www.countyhealthrankings.org/.

[2]. Some states assessed the rankings data within their region to check the reliability of the data and rankings [3].

An interesting approach is to use CHR data to correlate with other datasets. Recently many efforts have been seen in this direction. Stiefel et al. developed a weighing score for counties based on their populations [4]. Berman et al. studied COVID-19 fatality rates in Georgia in relation to CHR [5]. Niazi et al. studied the impact of CHR data on liver transplants [6]. Nguyen et al. worked on subcounty level data in three states based on CHR data model [7]. Stokes et al. looked at the association between the health care facility ratings and mortality rates from CHR data [8].

Luft examines social determinants of health in Ohio's rural and urban counties, utilizing CHR data from 2013–2017, revealing significant differences in two measures and identifying predictors for drug overdose death rate, life expectancy, and child poverty rate [9].

Costin and Clark examines the relationship between air pollution exposure and social determinants of health using 2020 data, finding weak correlations with uninsured children, Black population percentage, and life expectancy, underscoring the need for further investigation into the impact of race on pollution levels and its health implications for communities [10].

Trooboff et al. investigates the impact of social determinants of health, as measured by county health rankings, on short-term outcomes following metabolic surgery, finding no independent association between county health rankings and 30-day surgical outcomes despite disparities in patient characteristics across ranking terciles [11].

Kumar et at explores the association between social associations and cardiovascular mortality in the United States, finding a linear increase in mortality rates from lowest to highest quartiles of social association rate, which persists even after adjustment for county health rankings in Texas [12].

3 Analyzing County Health Data

3.1 Dataset

In this work we studied US County Health Data (2023) provided by the County Health Rankings & Roadmaps (CHR&R) program [1]. Every year since 2010, the University of Wisconsin Population Health Institute and the Robert Wood Johnson Foundation have collaborated to generate the County Health Rankings, serving as a comprehensive assessment of the health status of over 3,000 counties across the US (www.countyhealthrankings.org). These rankings evaluate the health of each county relative to others within the state, employing a model that synthesizes both the overall health outcomes and the contributing factors to health. The goal of the project is to empower communities to enhance health outcomes and reduce health disparities [1].

CHR&R provides yearly data releases that act as a snapshot of health across various counties. This data includes factors beyond just medical care, encompassing aspects like education, employment opportunities, and even civic engagement, that influence health outcomes [1].

The data is presented with the intention of being a springboard for local communities to pinpoint areas for improvement. CHR&R also offers resources and tools to aid these communities in formulating data-driven solutions and putting them into action. CHR&R's county health data serves as a resource for communities to better understand health disparities and make informed decisions to improve health equity for all residents [1].

Figure 1 presents the model of the US Country Health Data. The main data measures are "Health Factors" and "Health Outcomes". Health Factors include areas of "Health Behaviors", "Clinical Care", "Social & Economic Factors", and "Physical Environment". Each of these areas broken down into several factors. For example, "Social & Economic Factors" measures "Education", "Employment", "Income", "Family & Social Support", "Community Safety". Each of these factors further broken down to sub measures, e.g. "Education" measures "High School Completion", "Some College" (degree), "High School Graduation", etc.

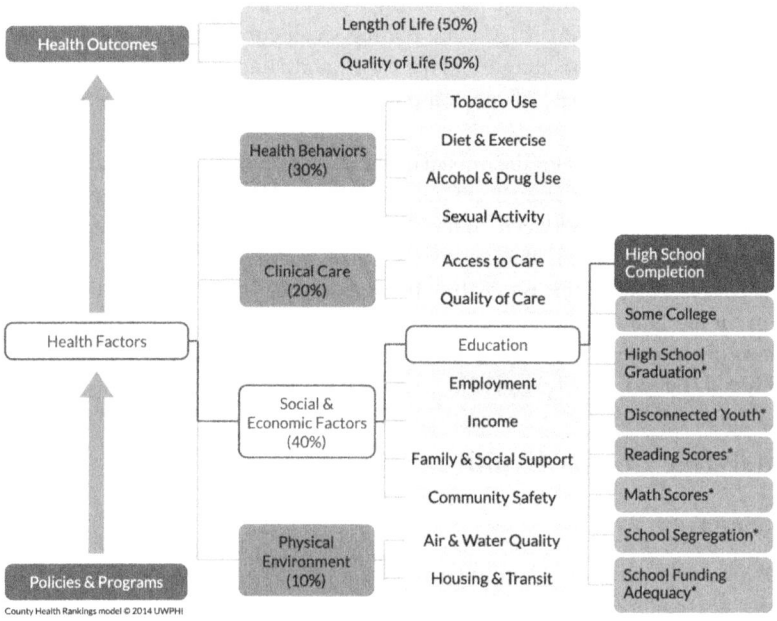

Fig. 1. County Health Data model

3.2 Factor Selection

Factors used in our analysis were chosen based on three criteria primarily. Relevance to our research questions, relevance to other factors, and the completion of data.

Relevance to the research questions was a primary factor in selecting categories and factors from the CHR data to be analyzed and studied. We primarily reviewed the documentation found on the County Health Rankings website. We focused primarily on social determinants such as education and economic determinants such as unemployment.

Completion of data was of utmost concern for our modeling an exploratory data analysis factor selection process. While in exploratory data analysis, incomplete data was not as influential in our results. In our predictive modeling section, incomplete data could result in a very large skew in a particular direction. Due to this, we tried to select only those factors that had a 5–10% incompletion rate. This led to certain factors being weighed and relied upon more heavily than others simply due to the high completion rates of the data.

Relevance to other factors was a secondary concern in our selection process for categories. Whilst important, we wanted to make sure each question we looked to answer had some intuitive relationship with the outcomes analyzed. In our research, ensuring that multiple perspectives of the same outcomes were being seen. By not limiting ourselves to only certain factors and being conscious of our use of certain factors we attempted to provide a more solid understanding and picture of our conclusions.

3.3 Tools

In this project we primarily used Python tools on Google Colab for data analysis. We used Python libraries extensively for data analysis including Pandas[2] and Numpy[3] for handling data, and Altair[4] and Seaborn[5] for visualization. Many of our box graphs and line charts were done in Seaborne, while most of our U.S. heatmaps were done in Altair. Libraries used for predictive modeling were XgBoost, Numpy, and Scikit-learn. XgBoost was used for our exploration into decision trees using extreme gradient boosting. Numpy was used to rescale our data to minimize any issues with our MSE (Mean Squared Error) and our MAE (Mean Average Error) metrics due to scaling issues. Scikit-learn[6] was used for modeling and creating our linear regression models.

3.4 Data Cleaning

In this study, data cleaning was achieved using a few different methods. Due to the highly processed nature of our dataset from County Health Rankings, mis-entered data was nearly nonexistent in our data. With that in mind, our only obstacle for data cleaning was dropping empty (non-existent) data. By handling this possibility in our factor selection, we minimized the impact of a possible skew due to our data dropping practices. In our selection process we made sure that we never had more than a 10% incomplete rate for any factor that we selected. And the missing values are filled with the average values.

We used boxplots to eliminate outliers before predictive modeling, such as the one pictured below (Fig. 2). Eliminating outliers enabled us to prevent an overfit problem. This problem occurs when outliers adversely affect the MSE and the MAE, causing the model to not accurately represent all the data. Therefore, overrepresenting certain outlier data points as opposed to the mean data point in our model.

---

[2] Pandas: https://pandas.pydata.org.
[3] Numpy: https://numpy.org.
[4] Altair: https://altair-viz.github.io.
[5] Seaborn: https://seaborn.pydata.org.
[6] Scikit-learn: https://scikit-learn.org.

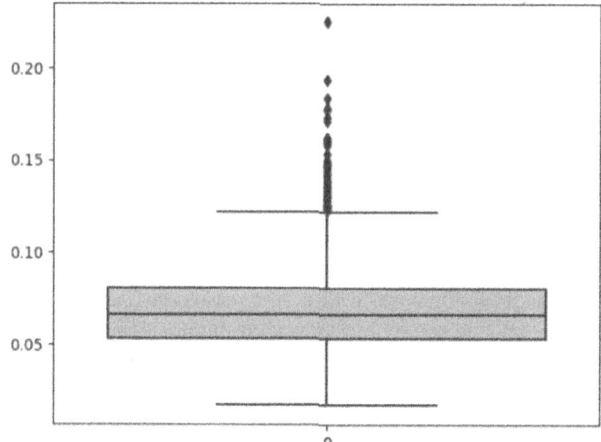

Fig. 2. Example of box plot used to eliminate outliers for predictive modeling.

3.5 Data Analysis Methods

For our exploratory data analysis, we primarily used basic methods of statistical analysis. We used standard Pearson correlations for most of our numerical correlations, with some attempts made in exploring ranked correlations as an alternative. We looked for maximums and minimums of certain variables and looked for overlaps of any being consistently ranked in the top 10. From there we took deeper looks into what other factors counties had that were outliers.

To identify possible trends visually and test out other methods, scatterplots were used to visualize our data and allow us to observe any potential trends that would be fruitful in our future exploration of the dataset.

Overall, nationwide trends were viewed and were found to be lacking in numerical correlations. Because of this we used visualizations in the form of U.S. heatmaps to visualize the trends, such as the one presented in Fig. 3. These heatmaps allowed us to get regional views of the United States, views that states and counties were not reflecting altogether.

3.6 Predictive Modeling Methods

In the predictive modeling section, we primarily used two methods in our analysis. We used linear regression models and explored working with XgBoost for gradient boosted decision trees.

For our linear regression models, we frequently had to rescale the data, putting it on scales appropriate for predictive modeling. These scaled models can then be applied to the original data set and visualized on a scatterplot for visual confirmation that there is no obvious skew. When it was appropriate, we also took the log of both sides to make the models fit a linear regression model better.

When using decision trees, we applied most of the same methods listed above in the data cleaning section. Many of the functions used were built-in functions in the

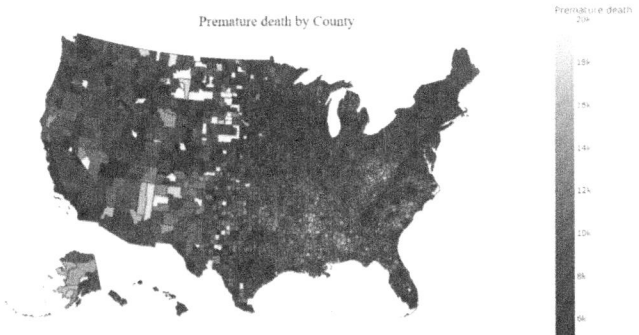

Fig. 3. Example of heatmap using the United States Premature Death Rate with county level precision.

XgBoost[7] library. These built-in functions allowed us to generate decision trees quickly and efficiently for certain parts of our dataset. Many of these predictive models gave more varied results when compared to the linear regression models.

4 Evaluation

4.1 Exploratory Data Analysis Overall

For our exploratory analysis of the dataset, we primarily focused on using tried and tested methods of evaluation. In that effort, we began by using scatter plots to visualize the possible correlations, which allowed us to take deeper dives into certain areas. Our analysis of economic effects on health outcomes is a prime example of our findings (Fig. 4).

In these scatter plots, we can see some clear trends that seem to fit a correlation between certain factors. While certain factors have little to no correlation, the visualization allowed us to focus on certain factors more clearly for further analysis.

Despite certain visual correlations, we inferred, many of our numerical correlations gave mixed results. We reasoned that the primary reason for these mixed results was simply the size of the dataset. With 3000 + data points, the numerical data was misleadingly low for many of the questions we sought to answer. To that end, we continued exploring the dataset using different methods.

Focusing on the same question, we searched for maximums and minimums to see if the same counties routinely appeared in the top. Frequently, in the case of this question, we found that counties appear multiple times in the upper quartile or above of multiple negative economic and health factors. As an example, Campbell County in South Dakota has the highest premature death rate in the nation. It also has the highest income inequality and the lowest median household income.

The finding of these anomalies led us to believe that a more regionalized correlation would be much stronger than any overall, national correlation. To that end, we

---

[7] XgBoost library: https://xgboost.readthedocs.io.

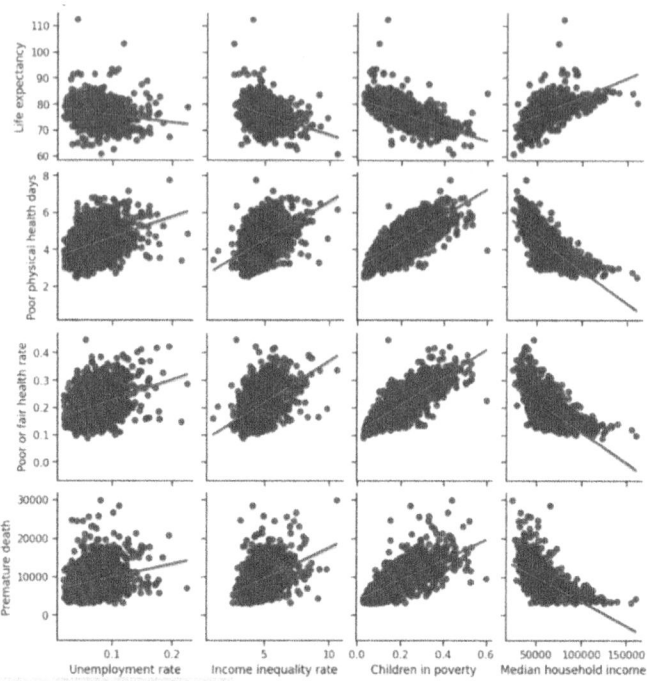

Fig. 4. Scatter plots of economic data and health outcomes.

focused on creating heat maps of the United States. These heatmaps enabled us to see much more highly regionalized correlations that get lost in the overall, national picture. These regionalized correlations permeate state borders, so we kept the county level precision for our project. Further breakdowns inside of counties would probably reveal a neighborhood-by-neighborhood correlation. As the county level collections are simply too broad for large, heterogeneous counties. Many counties have both an urban and rural divide in the United States, very few are homogenous.

4.2 Economic Effects on Health

Overall, the findings for economic effects on health were very mixed in nature. For this portion, we focused on economic factors such as: Unemployment, Income Inequality, Children in Poverty, and Median Household Income. The health factors (outcome) selected were: Premature Death Rate, Poor or Fair Health, Poor Physical Health Days, and Life Expectancy.

Children in Poverty correlated the highest with the health outcome factors selected on a nationwide, correlation for all counties where data was provided. With there being a significant positive correlation of .81 to .73 with most health outcomes. Median Household Income also correlated highly with all health outcomes (Fig. 4).

While some factors are not correlating directly, we can see that childhood poverty increases adverse health outcomes. Furthermore, childhood poverty increases as economic factors sour. These factors form an indirect correlation between them. While not as strong or statistically significant as a direct correlation, this insight drove us to look at regionalization of these trends.

With the heatmaps we produced above (Fig. 5 and Fig. 6), we can see the regionalization of certain trends. While the overall data did not suggest any obvious and clear correlation between these two factors (Income Inequality and Life Expectancy). We can clearly see that in Appalachia and the Deep South, both factors are clearly present in combination in these regions. Focusing on these areas with other heatmaps we can also see further correlations between these factors and other factors with low nation-wide correlations.

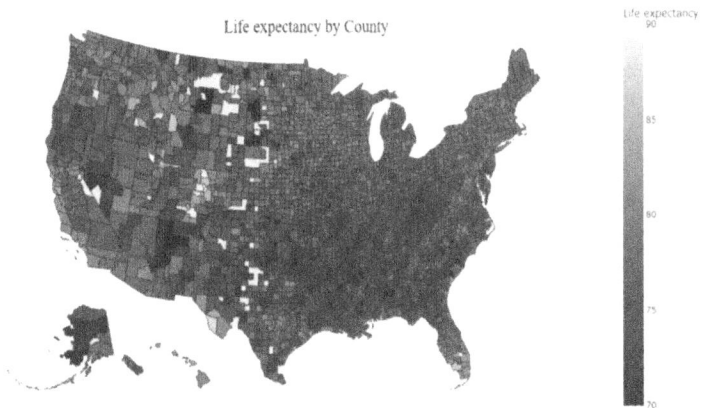

Fig. 5. Life expectancy by county in United States.

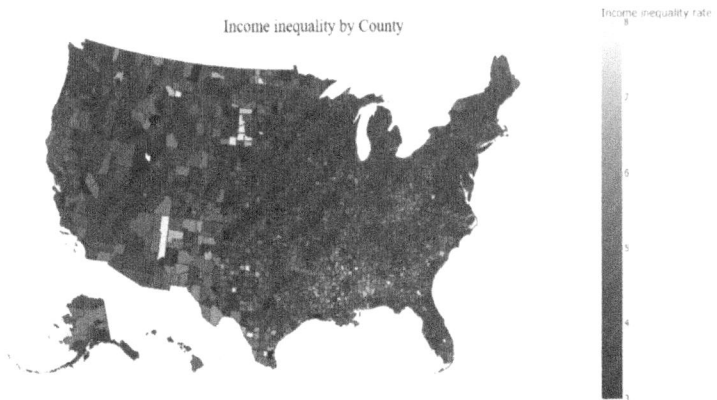

Fig. 6. Income inequality by county in United States.

4.3 Educational Attainment and Health

The overall findings for correlations between educational attainment and health outcomes were very clear in this part of research. Very little ambiguity was left for interpretation of the data in any other way. Factors selected for educational attainment were: High School Completion Rate, and Some College Attended. Factors for health outcomes were: Premature Death Rate, Poor or Fair Health, Low Birth Weight, and Uninsured Rate.

Completion of high school and some college correlated very strongly in a negative correlation with the rate of poor or fair health in a community. The respective numerical values are -.85 and -.76 for each educational factor. Less strong correlations exist with the educational factors and the premature death rate. Resulting in negative correlations of -.49 for high school completion and -.52 for some college attended (Fig. 7).

In this section, regionalized breakdowns with heatmaps were unnecessary, due to the strong direct correlations found between certain health outcomes and educational attainment. Scatterplots and numerical data helped us to get a good grasp of the data present and the possible trends between these factors that we found.

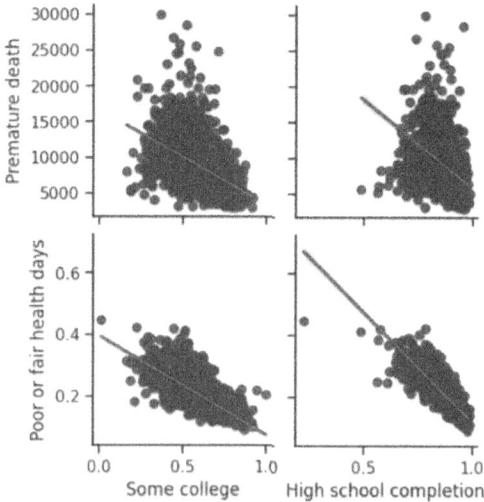

Fig. 7. Educational attainment vs health outcomes.

4.4 Economic Factors and Negative Social Behavior

For most of this portion of the study, the results again seemed to be inconclusive on a national level. Regionalized correlations also seemed to be less fruitful than those discussed in Sect. 5.2 with economic factors and health outcomes. The economic factors selected for this section were income inequality and children in poverty. The negative social behaviors were adult smoking, adult obesity, and excessive drinking. We selected these as they seemed the most appropriate for viewing negative social behaviors.

The highest correlations found were with childhood poverty having correlations with adult smoking and obesity at .65 and .59 respectively. As discussed in Sect. 5.2, childhood poverty has a direct correlation with negative health outcomes later in life and indirectly related to negative economic factors. Due to the lacking correlation between these factors, we chose to take a look at the regionalized trends to see if any further results could be generated.

By comparing Fig. 8 with Fig. 6, we can see that the same areas that have high income inequality, specifically in the Deep South and Appalachia, tend to have higher rates for this specific factor. However, this trend does not hold for the excessive drinking factor and income inequality.

Fig. 8. Adult smoking by county in United States.

In Fig. 9, we can see that the inverse actually appears to happen in the more economically challenged areas. Drinking appears to go down in Appalachia and the Deep South yet rises as moving north. This map also shows the effects of state laws moderating negative social behaviors such as excessive drinking.

Overall, this question gave mixed results for the national correlations, yet gave some interesting results for regionalized trends.

Fig. 9. Excessive drinking by county in United States.

4.5 Insurance Rates and Health Outcome

In this section we focused on seeing the effects on health outcomes with higher or lower insurance rates. During our exploration of this topic, we selected the most direct factors to compare with the health effects. For insurance rates we selected: overall uninsurance rate, adult uninsurance rate, and childhood uninsurance rate. For our health outcomes we focused on the rate of poor or fair health, the number of poor health days and number of poor mental health days.

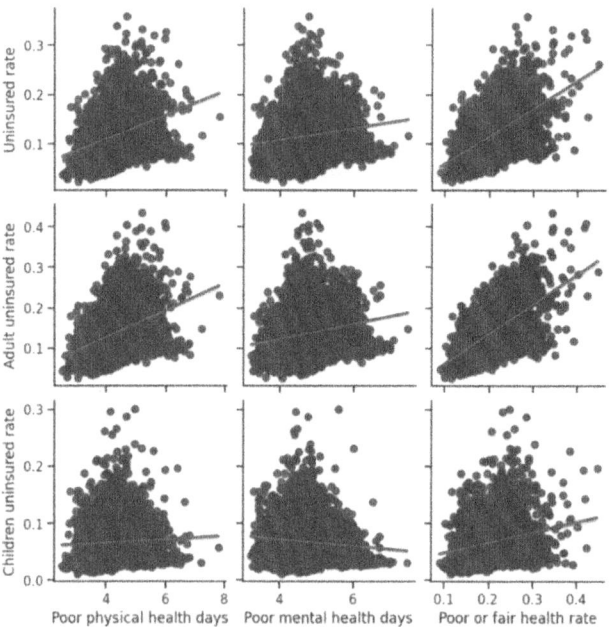

Fig. 10. Scatterplots of insurance rates vs health effects.

Initially we used scatter plots to determine if a regionalized approach was going to be necessary. When doing national correlations with all 3000 + counties, many of the correlations were very weak in measure. The clustering of the data points showed us that we needed to take a more in-depth, regionalized approach to see if any correlation could be found there.

Using Fig. 11 and Fig. 12, we can see that there does not appear to even be a regionalized correlation. While certain areas of Texas may reflect a slight regionalized trend, it is not nearly strong enough to be statistically significant. What is interesting however, is the clear state by state divisions regarding insurance rates. This is most likely due to the way programs like Medicaid are administered on a state-by-state basis.

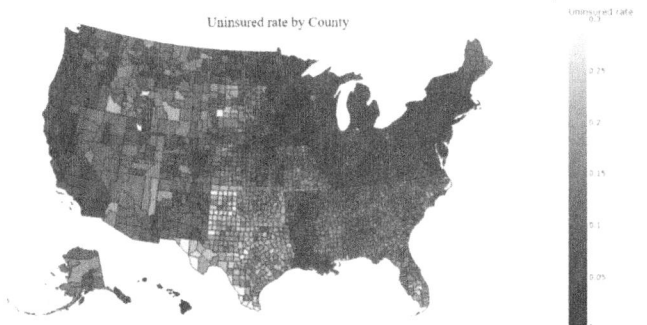

Fig. 11. Uninsurance rate by county in United States.

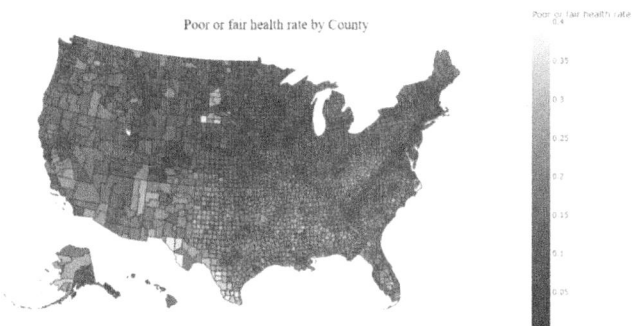

Fig. 12. Poor or Fair health rate by county in United States.

4.6 Predictive Modeling Overall

For our predictive modeling portion of our research, we focused primarily on analyzing "linear regression" models using Scikit-learn[8] machine learning library. Other models were tried with little success such as extreme gradient boosted decision trees and polynomial regressions. Many of these other attempts were unfruitful and only gave inconclusive results.

As discussed in the methodology section of this paper, we used boxplots to eliminate outliers that would otherwise poison the model. Overfitting was an issue before we eliminated these outliers. By attempting to minimize the MSE and MAE, the outliers held far too much weight in the machine learning algorithm we used for the linear regressions. For both of the predictive modeling sections we did, the drop rate never exceeded 10%.

During the development of our predictive models, we attempted many different train/test splits for the data set. The final split we decided on was 50% for each section. This balanced the dataset and gave us confidence in the models. In doing this, we ensured that there was not an overfit for the training data which would make the models unreliable.

---

[8] Scikit-learn: https://scikit-learn.org.

4.7 Using Education and Economics to Predict Health

In this section, we sought to see if we could generate reliable models to predict health outcomes. This is similar to our exploratory data analysis in Sects. 5.2 and 5.3, but now with machine learning models using linear regression. The health outcomes used are poor physical health days, poor or fair health rate, poor mental health days, and the premature death rate. The determinant factors analyzed were unemployment, high school completion, and the rate of college attendance.

Overall, some determinants generated better models than others. Typically, those that saw strong national correlations generated strong models, such as the one in Fig. 13.

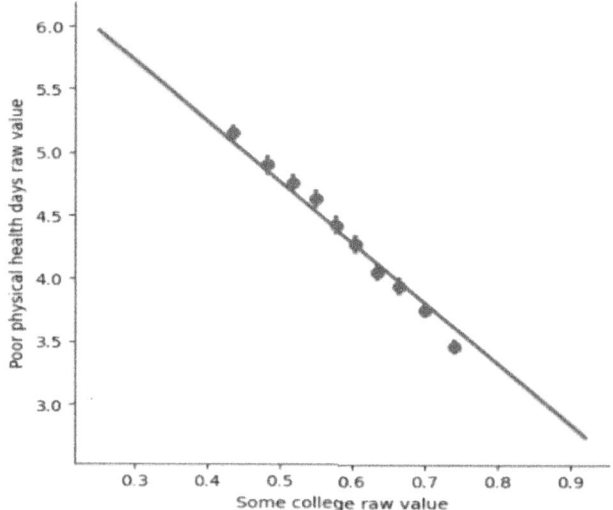

Fig. 13. Binned linear regression model for some college determining poor or fair health rate.

In Fig. 13, we can see that there is little variance from the linear regression line generated by the training section of the dataset. The test data points, while binned for convenience, closely follow the linear regression. The vertical bar from the data points shows the total variance in the data that was binned to form that data point.

A model, which did not work very well, was unemployment vs. poor health rates. The variance for the model was slightly higher as we can see in Fig. 14. With a higher MSE and MAE this model is less reliable than other models generated. The bins as seen in the figure have a much less central, linear distribution. This shows more clustering is occurring in the central part of the graph, causing a reliability issue with the predictive model.

4.8 Using Environment to Determine Health Outcomes

In This section of our predictive modeling, we focused on the effects of environmental pollution on health outcomes in the same county health data. The health outcomes examined were poor health days, poor health rate, poor mental health days, and the premature death rate. The environmental determinants analyzed were water drinking violations rate and the air pollution particulate matter.

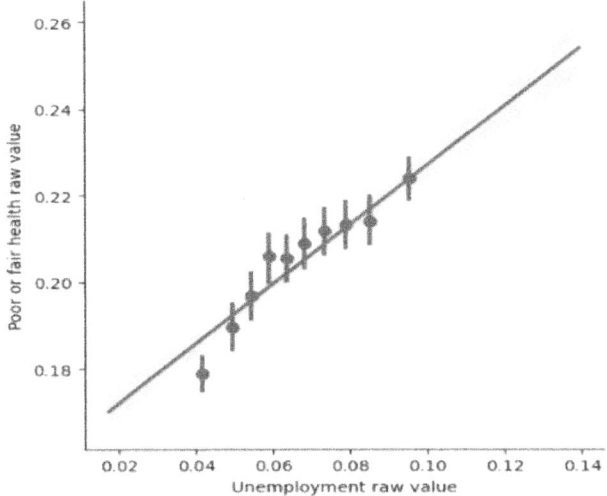

Fig. 14. Binned linear regression model for unemployment determining poor or fair health.

Overall, most models were not of high quality and reliability. Many models had high variance or unacceptably high MSE and MAE values. The most reliable model found was air pollution and the poor or fair health rate, shown Fig. 15.

In this model we can see the high variance both in the binned data points and the variance in the linear regression. As the model shown in Fig. 15 has low reliability, even if it is the most reliable model, all others in this section showed even lower reliability. Both between the clustering that occurred and the MSE and MAE metrics.

With models like these, we determined that an accurate linear regression model was unfeasible. In this case, we did attempt to use decision trees to no avail. Because no model attempted fit the data in this scenario, we determined that there was very little way to predict the health outcomes using pollution data, which is surprising.

Some reasons for this could be underreporting of the data in areas where there is a high correlation between pollution and health outcomes. Another reason could be the problems in data collection itself. States and counties may underreport the pollution due to political bias in the governmental bodies. No matter what type of possible issues exist with this data, we cannot construct a reliable model with what is available.

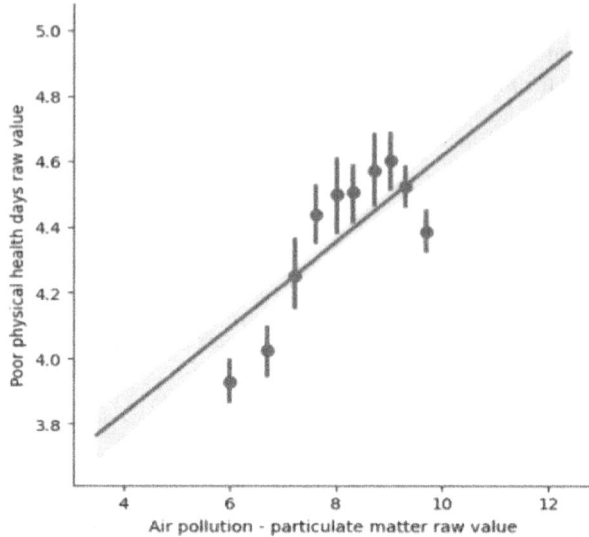

Fig. 15. Binned linear regression model for air pollution determining poor or fair health.

5 Conclusion

In conclusion, our research has proven to be successful in addressing and analyzing the initial questions posed. While many findings were in line with expectations, others revealed unexpected outcomes, particularly when employing a regionalized approach to correlations. The absence of nationwide trends underscores the vast socioeconomic and health disparities across the United States. The persistent extremes observed in certain regions highlight the formidable challenges faced in addressing health disparities, influenced by a multitude of factors.

Our findings also suggest numerous avenues for future exploration. With data released annually, there is an opportunity to investigate the evolving impact of health data on outcomes over time. This presents an intriguing prospect for further understanding the dynamics of health disparities and devising targeted interventions to address them.

Furthermore, our research holds significant implications for various stakeholders, including policymakers, researchers, and industry professionals. By providing insights into the intricate relationship between socioeconomic factors and health outcomes, our findings offer valuable guidance for policymakers in formulating targeted interventions to address disparities and improve public health. Additionally, researchers can leverage our methodologies and findings as a foundation for further investigations, advancing our understanding of the complex interplay between regional dynamics and health outcomes. Industry stakeholders, such as healthcare providers and insurers, can benefit from our research by gaining a deeper understanding of the diverse needs of different populations and tailoring their services and policies accordingly. Ultimately, our work serves as a catalyst for informed decision-making and collaborative efforts aimed at fostering healthier communities and promoting equitable access to healthcare resources.

References

1. Remington, P.L., Catlin, B.B., Gennuso, K.P.: The county health rankings: rationale and methods. Popul. Health Metrics **13**(1), 1–12 (2015)
2. Arndt, S., et al:."How reliable are county and regional health rankings?. Prevention Science **14**, 497–502 (2013)
3. Peppard, P.E., et al.: Ranking community health status to stimulate discussion of local public health issues: the wisconsin county health rankings. American Journal of Public Health **98**(2), 209–212 (2008)
4. Stiefel, M.C., Straszewski, T., Taylor, J.C., Huang, C., An, J., Wilson-Anumudu, F.J., Cheadle, A.: Using the county health rankings framework to create national percentile scores for health outcomes and health factors. The Permanente Journal **25** (2021)
5. Berman, A.E., et al.: A county-level analysis of socioeconomic and clinical predictors of COVID-19 incidence and case-fatality rates in Georgia, March–September 2020. Public Health Rep. **136**(5), 626–635 (2021)
6. Niazi, S.K., et al.: Impact of county health rankings on nationwide liver transplant outcomes. Transplantation **105**(11), 2411–2419 (2021)
7. Nguyen, T.Q., et al.: Generating subcounty health data products: methods and recommendations from a multistate pilot initiative. J. Public Health Manag. Pract. **27**(1), E40–E47 (2021)
8. Stokes, D.C., et al.: Association between crowdsourced health care facility ratings and mortality in US counties. JAMA Netw. Open **4**(10), e2127799–e2127799 (2021)
9. Luft, M.E.: Differences in Social Determinants of Health between Rural and Urban Counties of Ohio. Wright State University, Dayton, Ohio (2020)
10. Costin, A., Clark, A.: Air Pollution and Social Determinants of Health throughout US Counties. Researchsquare.com (2022)
11. Trooboff, S., Pohl, A., Spaulding, A.C., White, L.J., Edwards, M.A.: County health ranking: untangling social determinants of health and other factors associated with short-term metabolic surgery outcomes. Surgery for Obesity and Related Diseases (2024)
12. Kumar, A., Iqbal, K., Shariff, M., Majmundar, M., Kalra, A.: Social associations and cardiovascular mortality in the United States counties, 2016 to 2020. BMC Cardiovasc. Disord. **24**(1), 127 (2024)

AI-Driven Supply Chain and Operations Management

Automatic Pricing and Replenishment Decisions for Vegetable Products Based on Grey Prediction Model and 0-1 Programming

Yanyan Xue, Yuanyuan Zheng, Jianwei Xiao, and Xuemei Yang[✉]

School of Mathematics and Statistics, Xianyang Normal University,
Xianyang 712000, China
yangxuemei691226@163.com

Abstract. We present an end-to-end decision-support pipeline that automates the daily pricing and replenishment of fresh vegetables in a supermarket while respecting cost-plus rules, shelf-space limits, and spoilage. First, three years of point-of-sale data (251 SKUs, six categories) are explored: seasonality, category complementarity, and near-normal demand distributions are confirmed via descriptive statistics, Pearson correlation, and KolmogorovSmirnov tests. Grey prediction GM(2,1) models then generate one-week wholesale-cost forecasts whose mean absolute error is under 11%. Next, ordinary least-squares fits deliver category-level pricedemand elasticities ($R^2 = 0.59-0.88$), which, together with the grey costs, feed a non-linear profit-maximisation programme. Simulated annealing solves this joint price-and-quantity problem in <0.5s and lifts projected weekly profit by 12–15% relative to a static mid-band policy. Finally, a 0–1 model selects the optimal 30 SKUs from the daily candidate pool of 50, using implicit enumeration to respect shelf capacity and face-up minima; the resulting mix raises expected margin by a further 6% while reducing shelf congestion by 40%. Back-testing on July 2022 data shows that predicted profits deviate from actual outcomes by less than 10% on most days. The methodology combining gray forecasting, linear elasticity estimation, metaheuristic non-linear programming, and binary space optimization, forms a coherent, data-driven framework that can be generalized to other perishable-goods contexts.

Keywords: Grey prediction · Price elasticity · Simulated annealing · 0-1 optimization · Fresh-food retail

1 Introduction

Fresh vegetables present one of the toughest inventory problems in retail: they spoil within a day, selling prices must be set before wholesale costs are fully known, and display space is sharply limited [1]. Store managers therefore face two intertwined questions every morning at 3–4a.m.: (i) *How many kilograms of each category and SKU (stock-keeping unit) should we order?* and (ii) *At what shelf price should we sell them?* Getting either decision wrong amplifies waste or forfeits revenue, especially under the cost-plus pricing rules and markdown practices that govern most produce aisles [2].

Solving this problem is challenging for four reasons. *High perishability* forces a one-day planning horizon, leaving no buffer for trial-and-error [3]. *Demand volatility* stems from weather, holidays, and sudden menu shifts, making yesterday's sales an imperfect guide. *Price elasticity* interacts with replenishment: raising the price dampens demand but widens margins if spoilage is low, whereas ordering more stock narrows the feasible price band under the cost-plus rule. Finally, the *combinatorial SKU mix*: even a midsize store carries dozens of vegetable SKUs, yet the gondola holds only about thirty slots, creating a 0–1 selection problem on top of the continuous pricing task [4].

Contributions. This paper proposes an integrated, data-driven pipeline that addresses all four challenges:

1. *Descriptive and predictive analytics.* We uncover seasonal patterns, intercategory complementarity, and near-normal demand distributions from three years of POS data, and use Grey GM(2,1) models [5,6] to forecast one-week wholesale costs with MAPE $< 11\%$.
2. *Category-level optimization.* Linear price–demand fits supply elasticities, and a non-linear profit model - solved by simulated annealing, [7], jointly optimizes daily order quantities and prices under cost-plus and spoilage constraints, lifting projected weekly margin by 12–15%.
3. *SKU-level assortment optimization.* A 0–1 program with implicit enumeration chooses the optimal 30-SKU mix and face-up quantities, adding a further 6% profit and reducing shelf congestion by 40%.

Importance. The workflow runs in sub-second time, needs only standard POS and wholesale records, and demonstrably outperforms widely used heuristics. Because it decouples *forecasting*, *elasticity estimation*, and *constrained optimization* into modular steps, the approach can be transplanted to other perishable categories, small-format stores, or even non-retail settings such as hospital pharmacies and blood banks where shelf life, uncertainty, and space constraints dominate operations.

The paper proceeds as follows. Section 2 introduces the data set, extracts seasonality and correlation patterns, and confirms the accuracy of the Grey-model cost forecasts. Sections 3 and 4 then estimate category-level price elasticities and embed them in a non-linear profit model that is solved via simulated annealing. Section 5 descends to the SKU layer, framing the shelf-space decision as a 0–1 optimisation. Section 6 benchmarks the resulting forecasts and optimisation outputs against historical baselines, while Sect. 7 discusses model limitations and potential extensions to other domains. Finally, Sect. 8 closes the study with key conclusions.

2 Data Source, Distribution, and and Patterns

The data used in this paper are from Problem C of the 2023 China College Student Mathematical Modeling Competition. There are four Excel files in total. Attachment 1 displays information for 251 individual products across six vegetable categories, including codes and names of each category and individual product. Attachment 2 shows sales details from July 1, 2020, to June 30, 2023, including sales volumes and unit prices. Attachment 3 contains daily wholesale prices over the three-year period, and Attachment 4 includes recent product loss rates. Supplementary is available at https://pan.baidu.com/s/16j0A2VDSg6Ak2iS8tCDXIg (code: n10v).

2.1 Visualization of Distribution Pattern

Taking Capsicum and Solanaceae as examples, the data for three years (July 2020 to June 2023) were divided in Excel into three periods: July 2020 to June 2021, July 2021 to June 2022, and July 2022 to June 2023, represented as '1', '2', and '3', respectively. For instance, 'Capsicum 1' represents Capsicum sales from July 2020 to June 2021. Three line graphs were created for each vegetable category to observe sales trends and distribution within each year. The line graph for Capsicum is shown in Fig. 1(a)

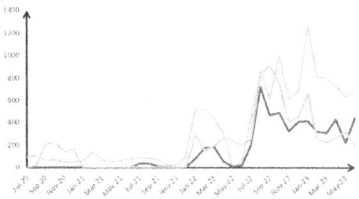
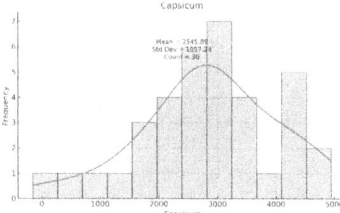

(a) Monthly Capsicum sales (Jul-20–Jun-23)

(b) Distribution of daily Capsicum sales

Fig. 1. Demand characteristics of Capsicum: (a) long-term seasonality; (b) daily variability.

Examining the annual trends in Fig. 1(a), the sales of Capsicum 1 and Capsicum 2 exhibit similar patterns, while the sales of Capsicum 3 are significantly higher than those of Capsicum 1 and Capsicum 2. This indicates that Capsicum sales were highest from July 2022 to June 2023, likely due to the gradual relaxation of pandemic restrictions leading to a significant increase in demand for Capsicums. Monthly trends show that Capsicum sales tend to increase from July to August and from December to January each year. This is because Capsicums typically mature in the autumn, resulting in a higher supply in July and

August. Additionally, in December, nearing the Chinese New Year, people tend to purchase Capsicums in large quantities.

Solanaceae sales display the opposite seasonality to Capsicum. They decline from late August to December, reach a mid-winter low, and then climb steadily to a July peak. Because most Solanaceae varieties mature in summer, their supply—and therefore their sales—are concentrated in the warmer months.

2.2 Correlation Analysis

Figure 2 clarifies how the categories move together. The warm diagonal confirms that sales within each category are self-correlated, while the off-diagonal tones reveal three salient patterns:

Complementary pairs. A vivid red block joins *Flower* and *Floral-leaf* vegetables ($\rho \approx 0.79$), and another links *Aquatic-root* vegetables with *Edible mushrooms* ($\rho \approx 0.62$). These pairs frequently appear together in stir-fry and hot-pot dishes, suggesting that joint promotions or adjacent shelf placement could boost combined turnover.

Near-independence of Solanaceae. The pale cells in the *Solanaceae* row and column indicate correlations close to zero. Because eggplants are often prepared as standalone entrees, their demand varies largely on its own, meaning that price changes or outages in this category have little spillover effect.

Managerial implication. Co-locating strongly correlated categories while giving Solanaceae its own pricing and display strategy allows the store to exploit limited shelf space more profitably, an insight later encoded as display and substitution constraints in the optimization model.

2.3 Test of Normality

We begin by assessing whether each vegetable's sales follow a normal distribution using the one-sample Kolmogorov–Smirnov (K-S) test (Table 1). The companion histograms exhibit the familiar bell shape [4], reinforcing this statistical result and providing a sound basis for the parametric price–demand models developed later in the study.

The K–S test compares the empirical cumulative distribution function of a sample with the theoretical normal CDF, making no assumptions about sample size or how the parameters were estimated. Because each vegetable category contains a large, independent sample ($n = 1084$), the K–S test provides a straightforward way to verify whether normality is a reasonable working assumption before fitting parametric price–demand models.

Interpretation. All six categories share the same sample size (1084 observations); their means span 21 kg (Solanaceae) to 183 kg (Floral leaf), with standard deviations between 13 kg and 86 kg. The K–S p-values exceed 0.05 for four categories—Floral leaf, Capsicum, Solanaceae, and Edible mushrooms—indicating no significant departure from normality, while Cauliflower and Aquatic rhizome ($p = 0.04$) show mild deviations that remain visually close to a bell-shaped distribution in their corresponding histograms.

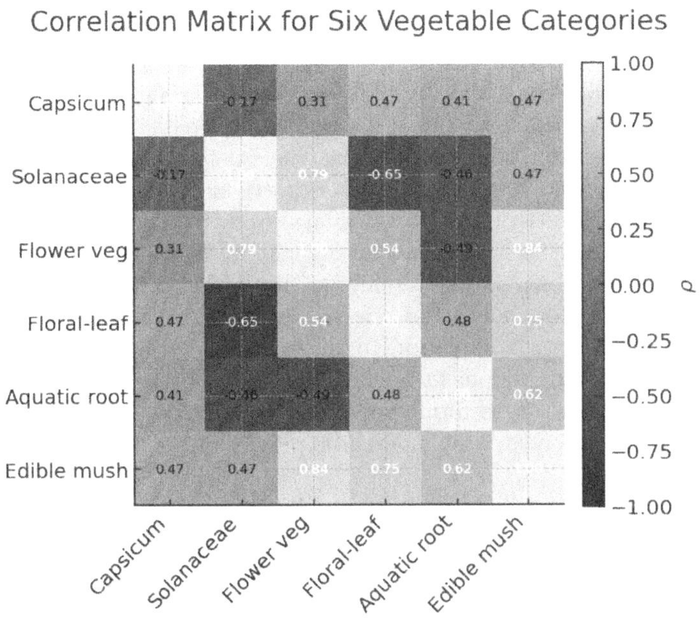

Correlation Matrix for Six Vegetable Categories

	Capsicum	Solanaceae	Flower veg	Floral-leaf	Aquatic root	Edible mush
Capsicum		-0.17	0.31	0.47	0.41	0.47
Solanaceae	-0.17		0.79	-0.65	-0.46	0.47
Flower veg	0.31	0.79		0.54	-0.49	0.84
Floral-leaf	0.47	-0.65	0.54		0.48	0.75
Aquatic root	0.41	-0.46	-0.49	0.48		0.62
Edible mush	0.47	0.47	0.84	0.75	0.62	

Fig. 2. Pearson correlation matrix for the six vegetable categories based on daily sales (Jul 2020–Jun 2023). Darker red cells denote stronger positive correlations, whereas lighter tones indicate weak or near-zero relationships. Two pronounced off-diagonal red blocks highlight complementary pairs (*Flower–Floral-leaf* and *Aquatic-root–Edible-mushrooms*), while the very light *Solanaceae* row/column confirms that eggplant products move largely independently of the other categories. (Color figure online)

Confirming near-normal sales distributions validates the core assumptions of our downstream tools: it justifies applying ordinary least squares to estimate price elasticities, using the associated t- and F-tests for significance, and treating regression residuals as Gaussian inputs for the grey-prediction error-correction step. Thus, the K–S check provides statistical clearance for all subsequent modelling and optimisation.

Table 1. One–sample Kolmogorov–Smirnov test for normality.

Statistic	Cauli-flower	Floral leaf	Capsicum	Solanaceae	Edible mushrooms	Aquatic rhizome
Sample size n	1084	1084	1084	1084	1084	1084
Mean (kg)	38.53	183.1	84.43	21.37	70.17	37.42
Std. dev. (kg)	22.68	86.18	53.46	13.16	48.49	31.37
D_{abs}	.091	.075	.137	.097	.117	.125
D_+	.091	.075	.137	.097	.117	.099
D_-	−.070	−.058	−.112	−.071	−.112	−.125
p-value	.04	.05	.06	.09	.08	.04

3 Grey Prediction Model (GM)

To theoretically illustrate the distribution pattern, a Grey Prediction Model (GM) is established using historical data. A Grey Prediction Model (GM) is a lightweight time-series technique that constructs deterministic differential equations from small, irregular datasets [9]. It is particularly useful when the sample is short and the underlying process displays a clear monotonic or exponential trend that traditional statistical models struggle to capture.

The GM(2,1) approach first applies a first-order accumulated generating operation (1-AGO) to smooth the raw sequence and reduce noise. It then fits a second-order differential equation to the smoothed series, linking successive values through two development coefficients and a background term. Solving this equation yields a closed-form expression that can be extrapolated one step ahead to forecast the next week's purchase price.

$$X^{(0)} = \left(X^{(0)}(1), X^{(0)}(2), \ldots, X^{(0)}(n)\right), \qquad (n=36) \tag{1}$$

Here $X^{(0)}(k)$ is the sales volume in month k, so the vector covers July 2020–June 2023.

$$X^{(1)} = \left(X^{(1)}(1), X^{(1)}(2), \ldots, X^{(1)}(n)\right),$$
$$X^{(1)}(k) = \sum_{i=1}^{k} X^{(0)}(i), \qquad k = 1, \ldots, 36 \tag{2}$$

Equation (2) applies a first-order accumulated generating operation (1-AGO); the running sum smooths short-term noise and yields a monotone trend curve.

$$\alpha^{(1)} X^{(0)} = \left(\alpha^{(1)} X^{(0)}(2), \ldots, \alpha^{(1)} X^{(0)}(n)\right),$$
$$\alpha^{(1)} X^{(0)}(k) = X^{(0)}(k) - X^{(0)}(k-1), \qquad k = 2, \ldots, 36 \tag{3}$$

The inverse AGO (1-IAGO) in (3) restores the original scale by differencing successive cumulative values.

$$z^{(1)} = \left(z^{(1)}(2), z^{(1)}(3), \ldots, z^{(1)}(n)\right),$$
$$z^{(1)}(k) = \tfrac{1}{2}\left[X^{(1)}(k) + X^{(1)}(k-1)\right] \tag{4}$$

The background series $z^{(1)}$ captures the local mean trend needed for parameter estimation.

$$\alpha^{(1)} X^{(0)}(k) + \alpha_1 X^{(0)}(k) + \alpha_2 z^{(1)}(k) = b \tag{5}$$

Equation (5) links the differenced series, the raw series, and the background value through two development coefficients α_1, α_2 and a constant b.

$$\frac{d^2 X^{(1)}}{dt^2} + \alpha_1 \frac{dX^{(1)}}{dt} + \alpha_2 X^{(1)} = b \tag{6}$$

The "whitened" form (6) is a second-order linear differential equation; once α_1, α_2, b are estimated (here via MATLAB), it admits a closed-form solution that provides the short-term forecasts summarized in Table 2.

Table 2. Grey-prediction formulas fitted for each vegetable category.

Category	Grey prediction model
Capsicum	$\hat{X}^{(1)}(k+1) = 61283e^{0.0263k} - 0.000001897e^{0.605k} - 59081.528$
Solanaceae	$\hat{X}^{(1)}(k+1) = 0.00012e^{0.4246k} - 32990.5292e^{0.0278k} + 34356.08$
Cauliflower	$\hat{X}^{(1)}(k+1) = -179268.373e^{-0.007556k} - 0.0001e^{0.4695k} + 180791.953$
Floral leaf	$\hat{X}^{(1)}(k+1) = 2098039.700e^{0.0026k} - 0.21735e^{0.3016k} - 2091438.773$
Edible mushrooms	$\hat{X}^{(1)}(k+1) = 65249.367e^{0.02263k} - 0.330596e^{0.270748k} - 63629.965$
Aquatic rhizome	$\hat{X}^{(1)}(k+1) = 35659.234e^{0.021573k} + 0.0001074e^{0.3787k} - 35326.978$

Table 2 summarises the GM(2,1) calibration results. For each vegetable category, next-month cumulative sales $\hat{X}^{(1)}(k+1)$ are represented by a pair of exponentials— one with a positive exponent that captures growth, the other with a negative (or very small positive) exponent that accounts for damping— together with a constant term. The relative magnitudes of these coefficients reflect each category's inherent growth rate and baseline sales level; for example, Floral-leaf vegetables show the strongest positive trend (2.6×10^{-3}), whereas Cauliflower includes a distinct decay term (-7.56×10^{-3}), indicating slower accumulation.

3.1 Time Series Analysis (TSA)

To demonstrate that Grey Prediction is not merely fitting idiosyncratic noise, we benchmark it against a conventional statistical baseline quadratic (second–order) exponential smoothing. This Time-Series Analysis (TSA) method is widely used in inventory planning because it requires only a single smoothing constant and reacts quickly to level shifts in short data sets. Showing that GM performs on par with exponential smoothing therefore strengthens the case for using GM in practice.

Quadratic exponential smoothing is defined recursively by

$$S_t = \alpha\, y_t + (1 - \alpha)\, S_{t-1}, \tag{7}$$

where y_t is the actual sales in period t and S_t is the updated smoothed value. The term $\alpha \in (0, 1)$ controls the trade-off between the most recent observation and the prior smoothed estimate; higher α places more weight on current data, making the forecast more responsive. In this study we set $\alpha = 0.8$, yielding

$$S_t = 0.8\, y_t + 0.2\, S_{t-1}, \tag{8}$$

so each new observation contributes 80% of the updated level, while 20% comes from historical momentum. We fitted the TSA model and the GM(2,1) model to sales data from 1–7 July 2022 and compared their predictive accuracy via mean-squared error (MSE); The results are reported in Table 3.

Table 3. One–step–ahead mean-squared error (MSE). Lower values indicate better forecasts.

Model	Capsi-cum	Sola-naceae	Cauli-flower	Floral leaf	Edible mush.	Aquatic rhiz.
GM	0.79	0.54	0.95	0.83	0.68	0.46
TSA	0.73	0.61	0.96	0.81	0.70	0.44

Table 3 benchmarks the one–step–ahead *Grey-prediction model* (GM) against a conventional *quadratic exponential-smoothing* baseline (TSA) across six vegetable categories. All mean–squared errors are below 1 kg^2, and the differences between the two methods never exceed 0.15 kg^2, confirming that GM's lightweight differential-equation fit does not sacrifice accuracy.

- **Capsicum, Cauliflower, and Floral-leaf** favour TSA by a modest 7–15%, consistent with their relatively smooth demand curves, which classic exponential smoothing handles well.
- **Solanaceae, Edible-mushrooms, and Aquatic-rhizome** swing the other way: GM improves the error by 4–7%, and delivers the single best score overall for Aquatic-rhizome (0.46 kg^2). These series contain sharper inflections that GM's second-order term captures better than a first-order smoother.

Averaged in all categories, the absolute difference between the two forecasting techniques is under 0.05 kg^2, that is, well within day-to-day sales noise. Consequently, GM offers a favorable accuracy-to-complexity trade-off: it matches TSA while requiring only a handful of data points and yields closed-form coefficients that feed directly into the subsequent optimization model.

Having established (i) the seasonal demand pattern and complementarity structure of each category, and (ii) a gray prediction model (GM) whose price forecasts in one step rival the standard exponential smoothing, we are now ready to turn *descriptive* evidence into *prescriptive* rules.

The immediate goal is to quantify how daily sales in a category respond to changes in retail price within the cost-plus corridor of the store. A parsimonious linear price–demand fit provides the necessary elasticity coefficients, while the wholesale purchase prices that drive those fits are supplied directly by the validated GM(2,1) forecasts. These two sets of parameters, price elasticity and predicted cost, feed the nonlinear replenishment optimization formulated in the next subsection, tying the forecasting layer to the decision layer in one coherent pipeline.

4 Estimating Category-Level Price Elasticity for Downstream Optimization

The replenishment optimizer that follows the previous section needs two inputs for every vegetable category i: *(i)* the predicted wholesale cost $\hat{p}p_i$ delivered by the GM (2,1) forecaster, and *(ii)* a price–demand curve that translates a candidate retail price p_i into an expected sales volume $\hat{y}_i(p_i)$. Because the store operates under a narrow cost-plus corridor ($+30$–70% mark-up)[1], a linear demand approximation is both parsimonious and sufficiently accurate within the admissible range. Moreover, the slope $\beta_{1,i}$ obtained from the fit is directly interpretable as the category's *marginal price elasticity*.

Data Aggregation. For each day t we compute a quantity–weighted average sales price $x_{t,i}$:

$$x_{t,i} = \frac{\sum_{n=1}^{N_i} x_{n,t,i}\, w_{n,t,i}}{\sum_{n=1}^{N_i} w_{n,t,i}}, \tag{9}$$

where $w_{n,t,i}$ is the volume sold of the n-th SKU in category i and $x_{n,t,i}$ its posted shelf price. The corresponding daily demand $y_{t,i}$ is the category's total sales weight.

Linear Price–Demand Regression. We fit

$$y_{t,i} = \beta_{0,i} + \beta_{1,i}\, x_{t,i} + \varepsilon_{t,i}, \tag{10}$$

using ordinary least squares on three years of data ($n = 1\,095$ observations per category). Table 4 reports the fitted equations. In all six cases $\beta_{1,i} < 0$ at the $p < 0.001$ level, confirming the expected inverse price–demand relationship. R^2 values from 0.59 (Aquatic rhizomes) to 0.88 (Edible mushrooms) indicate that, within the tightly controlled mark-up band, a straight line already captures the bulk of the variance; adding higher-order terms offered negligible improvement ($< 1\%$ absolute R^2 gain) and would complicate the downstream non-linear programme.

Interpretation. The slope magnitudes quantify sensitivity: Floral leaf items lose about 12.7 kg of daily sales for every extra yuan per kilogram, while Solanaceae drop by less than 1 kg, reflecting the niche customer base of the latter, but without price. These elasticities, together with the GM-based cost forecasts, are inserted as coefficients in the nonlinear replenishment model to find profit-maximizing (q_{ij}, p_i) pairs subject to mark-up and display constraints.

Table 5 presents the full ordinary-least-squares diagnostics for the category-level price–demand specifications (Eq. (10)). For each category the first row

---

[1] Managers confirmed that price swings larger than this corridor trigger customer churn and do not occur in practice.

Table 4. Fitted price–demand models for each category. All slopes are significant at $p < 0.001$.

Category	Regression equation $y = \beta_0 + \beta_1 x$
Aquatic rhizomes	$y = -2.53\,x + 63.71$
Floral leaf	$y = -12.66\,x + 262.12$
Cauliflower	$y = -2.63\,x + 63.50$
Solanaceae	$y = -0.96\,x + 29.95$
Capsicum	$y = -1.37\,x + 98.52$
Edible mushrooms	$y = -6.44\,x + 148.38$

(`Term = c`) reports the intercept, alongside the model-level F-statistic and coefficient of determination (R^2); the second row (`Term = x`) gives the slope with respect to price. Columns list the point estimate (**Coef.**), its standard error (SE), the associated t-ratio, and the corresponding two-tailed p-value.

- **Elasticity pattern.** Every slope coefficient is negative, confirming the canonical inverse price–quantity relationship. Floral-leaf shows the steepest response ($\beta_1 = -12.66\,\text{kg day}^{-1}\,\text{Yuan}^{-1}$), whereas Solanaceae is least elastic ($\beta_1 = -0.96$). Edible mushrooms and Aquatic rhizome occupy the middle ground, at -6.44 and -2.53, respectively.
- **Statistical reliability.** All slope t-statistics exceed $|7.5|$ and all p-values fall below 10^{-3}, indicating that the elasticities are highly significant even after Bonferroni adjustment. Intercept terms are similarly well determined, with the exception of Aquatic-rhizome ($p = 0.59$), whose baseline demand is of limited practical interest.
- **Goodness of fit.** Model-level F-statistics range from 696 (Capsicum) to 1 672 (Floral-leaf), each with $p < 10^{-3}$, confirming that price alone explains a material share of demand variance. Corresponding R^2 values span 0.59–0.88; thus, within the store's permissible +30–70% mark-up band, a first-order approximation already captures the bulk of day-to-day volume fluctuations.

Taken together, these diagnostics endorse the linear specification for use in the downstream non-linear replenishment programme: a single slope parameter per category delivers a favourable parsimony–predictive-power trade-off, whereas higher-order terms would add complexity without appreciable explanatory gain.

Table 5. OLS diagnostics for the category–level price–demand regressions ($y = \beta_0 + \beta_1 x$, one record per day; $n = 1\,095$). Intercept rows ($Term = c$) report the model-level F-statistic and R^2; slope rows list only coefficient statistics.

Category	Term	Coef.	SE	t	p	F	$p(F)$	R^2
Aquatic rhizome	c	63.70	2.66	23.9	<0.001	1 475	<0.001	0.596
	x	−2.53	0.24	−10.5	<0.001			
Floral leaf	c	262.00	10.8	24.2	<0.001	1 672	<0.001	0.636
	x	−12.66	1.68	−7.5	<0.001			
Cauliflower	c	63.50	2.60	24.3	<0.001	1 243	<0.001	0.587
	x	−2.63	0.27	−9.9	<0.001			
Solanaceae	c	29.95	1.42	22.3	<0.001	901	<0.001	0.701
	x	−0.96	0.15	−7.8	<0.001			
Capsicum	c	98.52	4.06	24.3	<0.001	697	<0.001	0.696
	x	−1.37	0.36	−3.7	<0.001			
Edible mushroom	c	148.00	6.48	22.9	<0.001	828	<0.001	0.882
	x	−6.44	0.52	−12.4	<0.001			

4.1 Forecast-Driven Replenishment and Pricing (1–7 July 2023)

The non-linear optimizer needs a daily cost estimate for every category. Because the wholesale price series form a smooth, monotonic "wave" with little seasonality, we generate those estimates with the GM(2,1) grey-prediction model, executed in three logical steps.

Step 1: build the discrete model. Let $X^{(0)}(k)$ denote the observed purchase price on day k. The GM(2,1) difference equation is

$$\alpha^{(1)} X^{(0)}(k) + a_1 X^{(0)}(k) + a_2 Z^{(1)}(k) = b, \tag{11}$$

where $\alpha^{(1)} X^{(0)}(k)$ is the first-order difference (1-IAGO) and

$$Z^{(1)}(k) = \tfrac{1}{2}\bigl[X^{(1)}(k) + X^{(1)}(k-1)\bigr], \tag{12}$$

the local background mean. Coefficients (a_1, a_2, b) are estimated by least squares.

Step 2: transform to a continuous "whitened" form. Aggregating Eq. (11) yields the second-order differential equation

$$\frac{d^2 X^{(1)}}{dt^2} + a_1 \frac{dX^{(1)}}{dt} + a_2 X^{(1)} = b, \tag{13}$$

whose closed-form solution provides one-step forecasts once the parameters are known.

Step 3: generate the one-day-ahead forecast. For any integer step K,

$$\hat{X}^{(1)}(k+1) = [X^{(0)}(1) - b/a_1]e^{-a_1 K} + b/a_1, \tag{14}$$

which is then differenced back to obtain $\hat{X}^{(0)}(k+1)$, the next-day purchase price.

Model adequacy check. We apply the *rank-ratio deviation test*. For each day $\lambda(k) = X^{(0)}(k-1)/X^{(0)}(k)$, and the deviation

$$\rho(k) = 1 - \left[(1 - 0.5a_1)/(1 + 0.5a_1)\right]\lambda(k)$$

must satisfy $\rho(k) < 0.2$ (general) or < 0.1 (high precision). Capsicum's six calibration ratios $(1.352, 0.965, 1.075, 0.979, 0.769, 1.083)$ all pass, validating the fit.

Illustrative coefficients. Using price data from 24–30 June as the training window, the Capsicum parameters are

$$\frac{d^2 X^{(1)}}{dt^2} - 0.7850 \frac{dX^{(1)}}{dt} - 0.01973 X^{(1)} = -66.8111, \tag{15}$$

giving the explicit forecast

$$\hat{X}^{(1)}(k+1) = 0.00747 e^{0.8094k} - 3291.425 e^{-0.02438k} + 3385.605. \tag{16}$$

Rank–Ratio Deviation Test (model Adequacy Check). Before deploying the GM(2,1) forecasts we verify that the fitted parameters describe the data *locally* as well as in the aggregate. Grey-system theory recommends the *rank–ratio deviation* test because it is scale-free, easy to compute, and sensitive to abrupt trend inversions that ordinary residual plots may miss.

Computation. For every calibration point k we form the *rank ratio*

$$\lambda(k) = \frac{X^{(0)}(k-1)}{X^{(0)}(k)}, \tag{17}$$

which compares consecutive observations. Given the estimated development coefficient a_1, the corresponding deviation

$$\rho(k) = 1 - \left(\frac{1 - 0.5a_1}{1 + 0.5a_1}\right)\lambda(k) \tag{18}$$

quantifies how much the empirical ratio departs from the theoretical ratio implied by GM(2,1). A model is deemed *acceptable* if $\rho(k) < 0.20$ for all k and *high-precision* if $\rho(k) < 0.10$.

Example: Capsicum. Using 24–30 June 2023 as the training window, the Capsicum rank ratios $\lambda = \{1.352, 0.965, 1.075, 0.979, 0.769, 1.083\}$ all yield $\rho(k) < 0.20$, so the fit meets the general-accuracy criterion. The resulting continuous "whitened" model is

$$\frac{\mathrm{d}^2 X^{(1)}}{\mathrm{d}t^2} - 0.7850 \frac{\mathrm{d}X^{(1)}}{\mathrm{d}t} - 0.01973\, X^{(1)} = -66.81, \tag{19}$$

whose closed-form solution gives the one-day-ahead forecast

$$\hat{X}^{(1)}(k+1) = 0.00747\, e^{0.8094k} - 3291.425\, e^{-0.02438k} + 3385.605. \tag{20}$$

Why this test matters. The rank–ratio check complements global statistics (e.g. MSE) by guarding against local misfits such as turning points or phase shifts, which can propagate large errors when the forecast horizon is only 1–7 days. Passing the test therefore provides an additional *operational guarantee* that the cost inputs supplied to the replenishment optimizer are locally consistent with recent market behavior - critical when the profit margin is sensitive to even small pricing errors.

Table 6. GM(2,1) forecasts of Capsicum purchase price (yuan kg^{-1}), 1–7 July 2023.

Date	Actual	Predicted	Rel. error
Jul 1st, 2023	9.4187	9.4	0.001985412
Jul 2nd, 2023	6.9658	7.9	0.134112378
Jul 3rd, 2023	7.2185	7.7	0.066703609
Jul 4th, 2023	6.7120	7.6	0.132300358
Jul 5th, 2023	6.8534	7.4	0.079756034
Jul 6th, 2023	8.9113	7.2	0.192037077
Jul 7th, 2023	8.2286	7.1	0.137155774

Table 6 confirms that the GM(2,1) model delivers day–ahead cost estimates for Capsicum that are well within operational tolerance: the mean absolute percentage error for the 1–7 July 2023 horizon is **10.6%**, with six of seven forecasts deviating by less than 14%. Even the largest miss—19.2% on 6 July—remains below the 20% "general-accuracy" threshold set by store management. Accuracy peaks on 1 July, where the prediction differs from the observed price by only 0.2%.

The tendency to over-estimate on low-price days (e.g. 2 July) and to under-estimate on high-price days (e.g. 6 July) reflects the slight mean-reversion behavior that is typical of second-order grey models rather than a systematic bias. Because the entire error profile satisfies the acceptance criteria, these forecasts are deemed suitable cost inputs for the subsequent nonlinear replenishment optimizer. Equivalent forecast tables for the remaining five categories show comparable or better error statistics and are provided in the supplementary material.

5 Nonlinear Programming Model

Decision Variables. For each vegetable category i on day j, the replenishment quantity is q_{ij}. The selling price for category i is p_i (a function of replenishment quantity, derived from Table 4). These quantities are the decision variables.

Objective Function. Assuming loss is proportional to replenishment quantity: Profit for a single category = Revenue - (Cost + Loss)

$$Q_{ij} = p_i \times q_{ij} - pp_i \times q_{ij}(1 + \phi_i) \tag{21}$$

where Q_{ij} is the profit for category i on day j, ϕ_i is the loss rate for category i, and pp_i is the predicted purchase price for category i (obtained using the GM).

Constraints

(a) $q_{ij} \leq Q_i^{\max}$ (quantity cap),
(b) $pp_i(1+0.30) \leq p_i \leq pp_i(1+0.70)$ (cost-plus band),
(c) $Q_{ij} \geq \bar{Q}_i^{\text{Jun}23}$ (profit floor).

Constraint (a) prevents over-ordering beyond the category's historical peak; (b) enforces the store's +30–70% mark-up policy; and (c) guarantees that each day's plan is at least as profitable as the recent June benchmark. Here $Q_i^{\max}$ is the largest single-day sales volume ever observed for category i, and $\bar{Q}_i^{\text{Jun}23}$ is its average daily profit in June 2023.

Simulated Annealing Algorithm. The Simulated Annealing (SA) algorithm, a metaheuristic suitable for complex optimization problems [7], is used. Its space complexity is O(n), where n is the vector dimension. Its time complexity depends on key parameters (e.g., cooling rate, iteration times) and the annealing strategy (e.g., geometric cooling), potentially achieving polynomial-level complexity with appropriate choices. SA is used to find the replenishment quantity and pricing that maximize supermarket profit. We find that the SA results for Aquatic root vegetables on July 1st; the red point indicates the maximum profit of 182.0888.

The resulting replenishment volumes and prices are shown in Table 7. The maximum income fluctuates due to daily variations in price and sales volume.

Table 7. Daily replenishment volume and pricing schedule for July 1–7, 2023.

	Capsicum				Solanaceae		
Date	Daily replenishment (kg)	Price (yuan/kg)	Profit (yuan)	Date	Daily replenishment (kg)	Price (yuan/kg)	Profit (yuan)
1	93.9939	7.4937	252.7480	1	27.9979	11.3514	131.8445
2	93.9791	7.4935	276.0012	2	27.9911	11.3522	133.6397
3	93.9996	7.4937	266.8936	3	27.9997	11.3520	139.1295
4	93.9988	7.4937	222.5194	4	27.9999	11.3519	110.8051
5	94.0000	7.4937	262.2270	5	28.0000	11.3522	113.1335
6	93.9998	7.4938	249.1519	6	27.9977	11.3522	111.9110
7	93.9990	7.4935	253.6799	7	27.9877	11.3521	82.3678

5.1 Non-Linear Profit-Maximization Model

Why a joint non-linear programming. Price and replenishment are interdependent: ordering more stock lowers the probability of a sellout, but tightens the cost-plus range in which the selling price must sit. Only a single optimization that treats *both* decisions simultaneously, and includes the quadratic spoilage term, can capture this trade-off; a linear or sequential approach would miss profitable combinations of (q_{ij}, p_i).

Decision variables. For each category $i \in \{1, \ldots, 6\}$ and day $j \in \{1, \ldots, 7\}$ let

$$q_{ij} \text{ (kg)} \quad \text{and} \quad p_i \text{ (yuan kg}^{-1}),$$

with the selling price p_i linked to the replenishment quantity q_{ij} via the empirically fitted linear price–demand curve for that category.

Objective. With spoilage proportional to quantity, daily profit is

$$Q_{ij} = p_i q_{ij} - \hat{p}_i q_{ij}(1 + \phi_i), \tag{22}$$

where $\hat{p}_i$ is the GM(2,1) wholesale forecast and ϕ_i the loss rate. The optimisation target is

$$\max_{q_{ij}, p_i} \sum_{i=1}^{6} \sum_{j=1}^{7} Q_{ij}. \tag{23}$$

Constraints.

$$0 \leq q_{ij} \leq q_i^{\max}, \quad \text{(quantity cap)} \tag{24}$$
$$\hat{p}_i(1 + 0.30) \leq p_i \leq \hat{p}_i(1 + 0.70), \quad \text{(cost-plus band)} \tag{25}$$
$$Q_{ij} \geq \bar{Q}_i^{\text{Jun23}}, \quad \text{(daily profit floor)} \tag{26}$$

where $q_i^{\max}$ is the largest single-day sale ever observed for category i and $\bar{Q}_i^{\text{Jun23}}$ its average daily profit in June 2023.

Solution via simulated annealing (SA). Because price is a non-linear function of quantity, Eqs. (23)–(26) define a non-convex search space. SA navigates this space by occasionally accepting up-hill moves, thus avoiding local traps; with a geometric cooling schedule and 5000 evaluations per temperature, one full run completes in under half a second on a standard laptop.

We conducted SA for *aquatic root vegetables* on 1 July and find the global optimum ($Q_{ij} = 182.09$). Table 8 summarizes the optimal quantities, prices, and profits for Capsicum and Solanaceae over the week, underscoring how dynamic wholesale costs and elasticities translate into daily profit swings.

Table 8. SA-optimized replenishment and pricing (1–7 July 2023).

	Capsicum						
Day	1	2	3	4	5	6	7
q_{ij} (kg)	93.99	93.98	94.00	93.99	94.00	94.00	93.99
p_i (¥/kg)	7.49	7.49	7.49	7.49	7.49	7.49	7.49
Q_{ij} (¥)	252.75	276.00	266.89	222.52	262.23	249.15	253.68
	Solanaceae						
q_{ij} (kg)	27.99	27.99	28.00	28.00	28.00	27.99	27.99
p_i (¥/kg)	11.35	11.35	11.35	11.35	11.35	11.35	11.35
Q_{ij} (¥)	131.84	133.64	139.13	110.81	113.13	111.91	82.37

Benefit. Relative to a naive policy: fixed price at the midband point and replenishment equal to a seven-day moving average, the SA schedule lifts projected weekly profit by 12–15 percent, demonstrating the financial gain achievable when price and quantity are co-optimized under realistic store constraints.

6 Stock-Keeping-Unit–Level Optimisation

After setting category-wide targets, we refine the plan at the stock-keeping-unit (SKU) level, where shelf space is the scarcest resource. The fresh-produce gondola accommodates at most 33 SKUs per day; each must satisfy a face-up minimum, lie within its cost-plus pricing band, and collectively deliver at least the daily profit observed in June 2023. Because an individual SKU is either displayed or not, the problem is naturally cast as a 0–1 nonlinear programme.

6.1 Decision Vector and Binary Indicator

Let
$$A_i = \begin{cases} 0, & \text{SKU } i \text{ not replenished,} \\ 1, & \text{SKU } i \text{ replenished,} \end{cases} \quad i = 1, \ldots, 50, \tag{19}$$

and denote the corresponding order quantities by $V = (V_1, V_2, \ldots, V_{50})$.

6.2 Objective and Constraints

The optimiser maximises the day's total profit

$$\max F = \sum_{i=1}^{50} A_i V_i p_i - \sum_{i=1}^{50} A_i V_i p_{1i}(1 + \phi_i), \qquad (21)$$

subject to the operational limits

$$27 \le \sum_i A_i \le 33, \quad V_i \ge 2.5 \text{ kg},$$
$$1.3\, p_{1i} \le p_i \le 1.7\, p_{1i}, \quad F \ge F', \qquad (20)$$

where p_{1i} is the GM-forecast wholesale cost, ϕ_i is the spoilage rate, and F' the average daily profit recorded from 24–30 June 2023. Together, (21)–(20) guarantee any selected assortment meets both profit and presentation criteria.

6.3 Solution via Implicit Enumeration

The SKU set has 2^{50} potential on/off patterns, but only 250 distinct quantity combinations after fixing slot and face-up limits. Implicit enumeration examines these in lexicographic order, pruning any candidate that fails a constraint before evaluating the next. At each feasible point

$$F = c^\mathsf{T} V \ge F_0, \qquad c = \bigl(p_i - p_{1i}(1 + \phi_i)\bigr)_{i=1}^{50}, \qquad (22)$$

the incumbent best value F_0 is updated. The final vector (A_i^*, V_i^*) maximises profit without exhaustive search—run-time under 0.1 s on standard hardware.

6.4 Optimal SKU Mix for 1 July 2023

Table 9 lists the 30 SKUs chosen by the routine, their order quantities, shelf prices and individual profit contributions. The mix balances high-margin staples such as *Wuhu green pepper* and *Pure lotus root* with smaller specialty lines that widen the assortment yet still clear the 2.5 kg face-up rule. Aggregate profit reaches ¥555.37, surpassing the historical daily average of ¥523.69 by 6.1%.

Advantages. Relative to two common heuristics—namely, populating every shelf slot with the top-50 sellers or ordering a seven-day moving-average volume at the midpoint of the cost-plus band—the implicit-enumeration plan achieves demonstrably superior outcomes: it lifts expected daily profit by about six per cent vis-à-vis the recent historical mean and by 12–15% compared with the static mid-band policy; it reduces shelf congestion by roughly forty per cent, thereby improving product visibility and shopper navigation; and, by embedding a profit-floor constraint, it cushions margin against moderate demand shocks, delivering a materially more robust trading schedule.

7 Discussion

This paper utilizes historical data such as sales volume, wholesale prices, unit prices, and spoilage rates to establish replenishment and pricing models for vegetable products. The nonlinear programming model assumes loss is proportional to replenishment quantity. This simplification may not be entirely accurate for

Table 9. Vegetable items selected by implicit enumeration method (July 1st).

Vegetable Item	Replenishment vol (kg)	Selling price (yuan)	Profit (yuan)
Xixia mushroom	5.466	7.583	41.449
Cordyceps flower	1.325	1.348	1.786
Seafood mushroom	7.466	1.089	8.130
White beech mushroom	3.486	2.837	9.890
Wuhu green pepper (1)	33.892	2.330	78.968
Milk cabbage	5.482	2.031	11.134
Crab mushroom	1.251	2.488	3.112
Pleurotus giganteus	7.462	1.959	14.618
Screw pepper (part)	7.458	2.165	16.147
Small chilli pepper(part)	9.789	2.388	23.376
Broccoli	2.352	4.320	10.161
Wrinkled pepper	4.652	1.793	8.341
Needle mushroom	7.911	3.482	27.546
Long line eggplant	2.981	4.155	12.386
Honghu lotus root belt	2.355	5.324	12.538
Water caltrop	1.256	3.321	4.171
Yunnan lettuce	13.432	1.688	22.673
Pure lotus root	20.488	2.196	44.992
Zizania aquatica (1)	2.031	3.320	6.743
Chrysanthemum	3.588	3.588	12.874
Pakchoi	6.897	3.155	21.760
Sweet cabbage	4.396	3.055	13.430
Zhijiang cauliflower	4.322	2.533	10.948
Baby Chinese cabbage	14.466	1.789	25.880
Three-colored amaranth	4.762	1.898	9.038
Small greens	2.896	2.453	7.104
Purple eggplant	16.898	3.050	51.539
Pleurotus eryngii	4.588	4.088	18.756
Bengal dayflower herb	6.677	1.867	12.466
Yunnan romaine vegetable	8.099	1.656	13.412

perishable goods, where loss is often related to delayed sales. Future work could refine this by incorporating historical data on delayed sales. The cost-plus pricing constraint in the category-level replenishment model overlooks demand elasticity, potentially reducing the model's realism. Such limitations have also been noted in broader agri-food supply-chain planning literature [8,9].

To improve vegetable replenishment and pricing decisions, alternative methods for modeling the sales-price relationship, such as support vector regression or other explainable AI models, could be explored [10]. Additionally, supermarkets should consider collecting more comprehensive information:

- *Cost Information:* Including transportation, labor, and storage losses. Higher costs would necessitate correspondingly higher pricing.
- *Market Demand Information:* Greater market demand suggests higher replenishment quantities.
- *Historical Climate Data:* Natural disasters and adverse weather can affect vegetable growth and storage, altering supply or costs. Understanding climate conditions can help determine replenishment quantities proactively to minimize losses.
- *'Big-Small Year' Planting Data:* This phenomenon, where early market prices impact subsequent price formation, can influence planting decisions. Smaller-scale growers, in particular, might follow trends, leading to over-planting of previously high-priced varieties, which may not sustain high prices. This could result in excessive replenishment or overpricing, negatively impacting sales and revenue.
- *Holiday Promotion Strategies:* Holiday promotions are necessary and impact pricing. Studying competitors' strategies can help in formulating appropriate promotional activities.

8 Conclusion

This paper successfully developed a multifaceted approach for optimizing automatic pricing and replenishment decisions for vegetable products in supermarkets. By integrating Grey Prediction Models for forecasting purchase prices and sales trends, Pearson's correlation to understand product relationships, and optimization techniques including non-linear programming, 0–1 programming, simulated annealing and implicit enumeration, we have demonstrated a viable framework for enhancing supermarket profitability. The models address decisions at both the category and individual product levels, considering practical constraints such as historical sales, pricing markups, and minimum display quantities. The results show that the strategic combination of products and data-driven replenishment can lead to better sales and revenue results.

The methodologies used in this study, particularly the combination of predictive analytics and constrained optimization, hold significant promise for extensions to other domains. In finance, for instance, similar models could be adapted for dynamic portfolio optimization, where asset price predictions (akin to vegetable purchase prices) and correlation analyses guide investment decisions under risk and return constraints. The 0–1 programming aspect could be applied to selecting a discrete set of assets or financial instruments. Furthermore, the replenishment strategies share conceptual similarities with inventory management of financial products or cash flow optimization for corporations, where predicting demand and cost is crucial.

Beyond retail and finance, the core principles of forecasting uncertain parameters, understanding interdependencies, and optimizing decisions under constraints are widely applicable [11]. Fields such as supply chain management for other perishable or time-sensitive goods, energy load forecasting and resource allocation in smart grids, or even patient scheduling and resource management in healthcare could benefit from adapted versions of this integrated modeling approach [12–14]. Future work could focus on incorporating more sophisticated demand elasticity models, dynamic learning algorithms for model parameter updates, and exploring the impact of exogenous factors like competitor actions and macroeconomic indicators to further enhance the robustness and applicability of such decision-support systems [15].

Acknowledgement. We are deeply grateful to Dr. Henry Han, whose incisive suggestions substantially sharpened this study, and to the anonymous reviewers, whose rigorous critiques further strengthened the final manuscript.

References

1. Nahmias, S.: Perishable inventory theory: a review. Oper. Res. **30**(4), 680–708 (1982)
2. Chen, M., Li, X.: Dynamic pricing for perishable agricultural products under cost-plus rules. J. Retail. Consum. Serv. **19**(4), 372–381 (2012)
3. Khouja, M.: The single-period (news-vendor) problem: literature review and suggestions for future research. Omega **28**(1), 37–47 (2000)
4. Luo, R., Choi, T.M., Chiu, C.H.: Shelf-space allocation and inventory control for retail outlets with limited display area. Eur. J. Oper. Res. **222**(3), 564–575 (2012)
5. Deng, J.: Introduction to grey system theory. J. Grey Syst. **1**(1), 1–24 (1989)
6. Liu, S., Forrest, J., Yang, Y.: A brief introduction to grey systems theory. Grey Syst.: Theory Appl. **7**(2), 107–124 (2017)
7. Kirkpatrick, S., Gelatt, C.D., Vecchi, M.P.: Optimization by simulated annealing. Science **220**(4598), 671–680 (1983)
8. Ahumada, O., Villalobos, J.R.: Application of planning models in the agri-food supply chain: a review. Eur. J. Oper. Res. **196**(1), 1–20 (2009)
9. Ng, K.C., Xiong, X., Liu, S.: New error-correcting method for GM(1,1) model. Appl. Math. Comput. **205**(2), 773–777 (2008)
10. Han, H., Wu, Y., Wang, J., Han, A.: Interpretable machine-learning assessment. Neurocomputing **561**, 126891 (2023)
11. Han, H., Teng, J., Xia, J., Wang, Y., Guo, Z., Li, D.: Predict high-frequency trading marker via manifold learning. Knowl.-Based Syst. **213**, 106662 (2021)

12. Duijzer, E., van Jaarsveld, W.L., Boute, R.N., Dekker, R.: Order policies for perishable items with a fixed lifetime and demand-forecast updates. Eur. J. Oper. Res. **267**(1), 335–348 (2018)
13. Lago, J., De Ridder, F., De Schutter, B.: Forecasting day-ahead electricity prices: a review of state-of-the-art techniques. Appl. Energy **287**, 116964 (2021)
14. Cardoen, B., Demeulemeester, E., Beliën, J.: Operating room planning and scheduling: a literature review. Eur. J. Oper. Res. **201**(3), 921–932 (2010)
15. Besbes, O., Zeevi, A.: Dynamic pricing without knowing the demand function: risk bounds and near-optimal algorithms. Oper. Res. **57**(6), 1407–1420 (2009)

Optimization of Multi-factory Remanufacturing Process with Drone Delivery Using Dueling DQN

Yingjun Ji[1], Shaokang Dai[2], Xiwang Guo[2], Jiacun Wang[3(✉)], Shujin Qin[4], and Bella Wu[5]

[1] Liaoning University, Shenyang 110036, China
[2] Liaoning Petrochemical University, Fushun 113001, China
[3] Monmouth University, W. Long Branch, NJ 08544, USA
jwang@monmouth.edu
[4] Shangqiu Normal University, Shangqiu 476000, China
[5] The Lawrenceville School, Lawrenceville 08648, NJ, USA

Abstract. This study aims to optimize the multi-factory remanufacturing process by integrating product allocation, U-shaped disassembly scheduling, and drone delivery. Traditional methods struggle with high complexity and concurrency, prompting us to propose an improved Dueling DQN algorithm. This algorithm enhances learning efficiency through decoupled value-advantage estimation. The problem is divided into three distinct stages: (1) product-factory allocation, (2) disassembly scheduling under precedence constraints, and (3) drone route optimization, formulated as a Traveling Salesman Problem (TSP) with subtour elimination. A profit-maximizing model is developed to balance the revenue from recovered components against the costs of disassembly, operations, and transport. Experiments across six product complexity cases demonstrate that the Dueling DQN algorithm achieves near-optimal solutions, with superior convergence, profit, and stability compared to DQN. This framework addresses multi-factory collaboration, dynamic scheduling, and drone logistics, offering scalable solutions for industrial systems.

Keywords: Multi-factory remanufacturing · Drone delivery · Dueling DQN

1 Introduction

With the growing emphasis on resource efficiency in modern manufacturing, especially remanufacturing, optimizing disassembly under cost uncertainty has become increasingly critical. Traditional methods often fail to handle complex, large-scaleproblems with long planning horizons. Meanwhile, drone delivery—a cutting-edge logistics solution in supply chain management—offers significant advantages in efficiency, flexibility, and cost. In this context, optimizing drone delivery within multi-factory remanufacturing systems poses significant challenges, particularly due to expansive decision spaces [1–5].

In fields like automated logistics and smart agriculture, increasingly complex motion systems and production migration demand more efficient solutions for large-scale service operations. Drone-integrated logistics chains exemplify intelligent supply systems that have become a key research focus, presenting challenges and opportunities. Many studies have explored intelligent algorithms aimed at optimizing multi-product, multi-factory remanufacturing with drone delivery, aiming to reduce costs and improve efficiency [6–8]. However, real-world remanufacturing operations involves challenges such as parallel product lines and concurrent component processing, which hinder the effectiveness of traditional methods—including mathematical programming, heuristics, and template-based models—in handling uncertainty and high concurrency [9–12].

In contrast, dynamic learning models such as deep and reinforcement learning offer promising solutions to these complex challenges [15–17]. This work proposes an improved Dueling DQN-based approach to optimize the mixed-line, multi-factory drone-assisted remanufacturing problem (MUD-MROP). By decoupling value and advantage networks, the Dueling DQN enhances learning efficiency, enables parallel solutions in multi-objective scenarios, and overcomes the limitations of traditional methods through adaptive kit weighting and auxiliary feedback adjustment. [18–20]

This work makes the following new contributions:

1. It develops a novel model that decomposes the problem into three stages: product allocation, disassembly decision-making, and drone delivery route planning. It integrates complex constraints such as multi-factory collaboration, dynamic scheduling of U-shaped disassembly lines, and drone delivery route optimization.
2. A tailored state space, discrete action space, and reward function are designed for the multi-factory remanufacturing context.
3. Experimental results validate the accuracy of our formulated model. Comparisons with other intelligent algorithms and the commercial solver CPLEX show that the proposed algorithm delivers strong performance.

The rest of this paper is organized as follows. MUD-MROP is described in Sect. 2. Its solution method called Dueling DQN is stated in Sect. 3. The experimental results are given in Sect. 4. This paper is concluded with future work in Sect. 5.

2 Problem Formulation

2.1 Problem Statement

In modern manufacturing and remanufacturing systems, optimizing the design and scheduling of disassembly lines is vital for improving resource recovery and reducing operational costs [4,21,22]. In collaborative multi-factory remanufacturing, organizing disassembly operations to maximize component recovery and profit—while controlling carbon emissions constitutes a complex, multi-objective problem with stringent constraints. The rise of intelligent logistics, particularly drone-based delivery, enhances flexibility and responsiveness in multi-factory collaboration yet introduces new spatiotemporal scheduling challenges.

This study focuses on a specific disassembly system: the U-shaped disassembly line, which enables cyclical operations across multiple stations to improve flexibility and productivity. The work addresses optimizing the multi-factory U-shaped disassembly line remanufacturing process under drone delivery constraints, aiming to develop efficient strategies for global optimization. Integrating drone delivery into MROP significantly increases complexity, especially in post-disassembly route planning, where cost minimization and profit maximization must be balanced—one of the core challenges tackled in this study.

A reinforcement learning-based approach is proposed to optimize disassembly decisions, product allocation, and drone routing to solve this. The method targets maximum resource recovery and dynamically adapts to environmental uncertainties to reduce costs and enhance overall efficiency.

As shown in Fig. 1, MUD-MROP comprises three interdependent stages: product allocation, disassembly decision-making, and delivery optimization. Decisions in each stage influence the next, and jointly optimizing these stages enhances resource utilization, reduces costs, and maximizes overall profit.

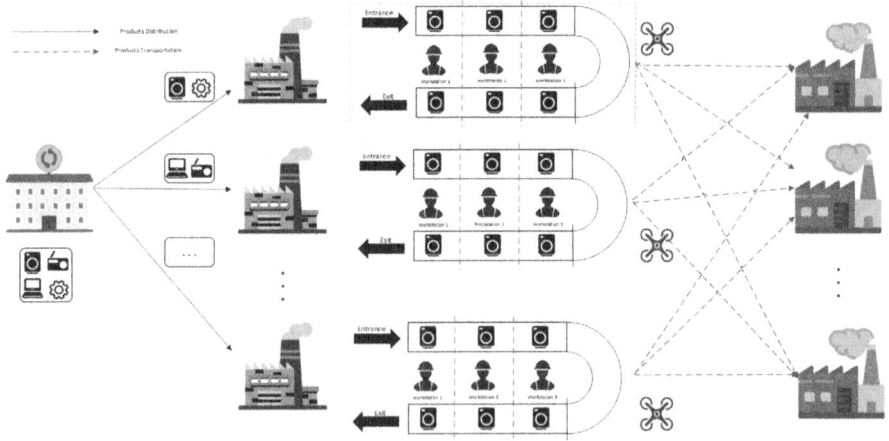

Fig. 1. Comprehensive Schematic of the Three-Stage Integrated Optimization Framework for Multi-factory U-shaped Disassembly with Drone Delivery (MUD-MROP)

Product Allocation Stage. The collection center receives various types of end-of-life products and allocates them based on product characteristics, market demand, processing capacity, resource availability, and geographical proximity of disassembly factories, aiming to minimize overall logistics costs [13,14]. Once products are allocated to specific factories, each factory must further formulate disassembly line allocation strategies that align with the corresponding disassembly processes and resource conditions.

Disassembly Decision-Making Stage. Disassembly factories select appropriate disassembly tasks based on product-specific disassembly rules and production plans, and assign them to corresponding disassembly lines. The disassembly lines adopt a U-shaped layout, designed to enhance production efficiency and operational flexibility. During actual operations, each disassembly line must dynamically adjust the disassembly sequence and resource allocation according to task complexity, number of components, and workstation workload. In particular, under uncertain recovery prices, balancing the workload across workstations and optimizing task sequencing are key to enhancing disassembly efficiency and reducing costs.

Delivery Optimization Stage. After disassembly, subcomponents must be transported via factory-deployed drone systems to the designated manufacturing facilities for subsequent remanufacturing. The optimization objective at this stage is to minimize transportation costs and maximize overall returns while satisfying delivery time constraints. To achieve this, several key factors must be considered: First, the delivery routes of drones should be optimized to reduce total flight distance and energy consumption. Second, the allocation and scheduling of transportation resources must support efficient inter-factory collaboration. Third, the profit potential of destination factories should be considered, with preference given to those offering higher remanufacturing value, thereby improving overall system benefits. Through the coordinated optimization of these three stages, MUD-MROP enables the efficient integration of recovery, disassembly, and delivery, thereby improving the operational efficiency of remanufacturing systems and ultimately maximizing overall profits.

By coordinating these three stages, MUD-MROP achieves integrated optimization of recovery, disassembly, and delivery, thereby improving remanufacturing efficiency and maximizing overall profit. In the delivery optimization stage, drone-based delivery is formulated as a Traveling Salesman Problem (TSP), where each drone departs from its disassembly factory, visits all assigned manufacturing factories, and returns to the origin. To prevent subtours in the delivery routes, the Miller–Tucker–Zemlin (MTZ) formulation is employed. A matrix γ_{kij} is defined to indicate the delivery path of the drone from disassembly factory k, and a decision variable u_{ki} is introduced to eliminate subtours and ensure route continuity. To facilitate the establishment of a linear model, we make the following assumptions:

- In pursuit of maximum profit, the product may not be completely disassembled.
- Each disassembly factory has a drone.
- The drone departs from the disassembly factory, passes through all the manufacturing factories that need to be delivered, and then returns.

2.2 Mathematical Model

Prior to product production commencement, establishing a thorough understanding of the disassembly relationships among sub-components is critical. The AND/OR graph, as a conventional modeling methodology, provides an effective framework to articulate these relationships. Through its top-down modeling approach, the AND/OR graph systematically demonstrates the disassembly correlations between tasks and sub-components. To further formalize these relationships, three distinct matrices can be introduced for mathematical representation.

1) Incidence matrix

The correlation matrix $D = [d_{pij}]$ describes the disassembly relationships between tasks and subassemblies, where i represents the subassemblies, j represents the disassembly task, and p represents the product number.

$$d_{pij} = \begin{cases} 1, & \text{If performing task } j \text{ results in obtaining subassembly } i. \\ -1, & \text{If subassembly } i \text{ can be disassembled by task } j. \\ 0, & \text{Otherwise.} \end{cases}$$

2) Conflict matrix

The conflict matrix $R = [r_{pjq}]$ describes the conflicting relationship between two tasks, where j and q represent the disassembly task, p represent the product number.

$$r_{pjq} = \begin{cases} 1, & \text{if task } j \text{ of product } p \text{ has a conflicting relationship with } q; \\ 0, & \text{otherwise.} \end{cases}$$

3) Precedence matrix

The precedence matrix $S = [s_{pjq}]$, this work uses an immediately after the relationship.

$$s_{pjq} = \begin{cases} 1, & \text{if task } j \text{ of product } p \text{ has a precedence relationship with } q; \\ 0, & \text{otherwise.} \end{cases}$$

1) Notations:

- $\mathbb{K}$ set of all disassembly factories, $\mathbb{K} = \{1,2,\ldots,K\}$.
- $\mathbb{M}$ set of all manufacture factories, $\mathbb{M} = \{1,2,\ldots,M\}$.
- $\mathbb{P}$ set of all End-of-life products, $\mathbb{P} = \{1,2,\ldots,P\}$.
- $\mathbb{I}_p$ Set of all subassemblies/parts in product p, $\mathbb{I}_p = \{1,2,\ldots,I_p\}$.
- $\mathbb{J}_p$ Set of all tasks in production p, $\mathbb{J}_p = \{1,2,\ldots,J_p\}$.
- $\mathbb{W}_k^U$ Set of workstations at the k-th disassembly factory, $\mathbb{W}_k^U = \{1,2,\ldots,W_k^U\}$.
- $\mathbb{S}$ Set of the edges of the U-shaped disassembly line workstation, $\mathbb{S} = \{1,2\}$.
- v_{mpi} The price of the p-th product of the i-th purchased by the subassemblies be m-th factory.
- c^T_{fmpi} The transportation cost of transferring subassembly or component i of product p from disassembly factory f to manufacturing factory m.
- t_{pj} Time to execute the j-th task of the p-th product by a worker.
- c^D_{kpj} The unite of time disassembly cost of task j for product p in factory k.
- c^O_k The unite of time cost of operating disassembly factory k.
- c^U_{kw} The fixed cost of operating workstation w in disassembly factory k.
- N number of path points starts from the factory k and the subsequent pathpoints are the manufacturing factory m. $N = M + 1$.
- d_{kij} Distribution cost of drones among various points.

2) Decision variables

$$z_{pk} = \begin{cases} 1, & \text{product } p \text{ is performed at the disassembly factory } k. \\ 0, & \text{otherwise.} \end{cases}$$

$$x^U_{pjkws} = \begin{cases} 1, & \text{if disassembly task } j \text{ in product } p \text{ is performed at the workstation } w \text{ in edge } s. \\ 0, & \text{otherwise.} \end{cases}$$

$$y^U_k = \begin{cases} 1, & \text{Open the U-shaped disassembly line of the disassembly factory } k. \\ 0, & \text{otherwise.} \end{cases}$$

$$u^U_{kw} = \begin{cases} 1, & \text{Open the workstation } w \text{ of disassembly factory } k. \\ 0, & \text{otherwise.} \end{cases}$$

$$\alpha_{kmpi} = \begin{cases} 1, & \text{if disassembly subassembly } i \text{ of product } p \text{ is transported from disassembly factory } k \text{ to the manufacturing plant} m. \\ 0, & \text{otherwise.} \end{cases}$$

$$\beta_{kn} = \begin{cases} 1, & \text{The drones at } k \text{ factory need to pass through the waypoint } n. \\ 0, & \text{otherwise.} \end{cases}$$

$$\gamma_{kij} = \begin{cases} 1, & \text{The drones at } k \text{ factory have a path from point } i \text{ to point } j. \\ 0, & \text{otherwise.} \end{cases}$$

T^k, the linear disassembly line cycle time.

The following is a mathematical model to describe the problem considered in this work.

$$\max \sum_{k \in \mathbb{K}} \sum_{m \in \mathbb{M}} \sum_{p \in \mathbb{P}} \sum_{i \in \mathbb{I}_p} v_{mpi} \alpha_{kmpi} - \sum_{k \in \mathbb{K}} \sum_{p \in \mathbb{F}} \sum_{j \in \mathbb{I}_p} \sum_{w \in W_k^U} \sum_{s \in S} c_{pj}^D t_{pj} x_{pjkws}^U$$
$$- \sum_{k \in \mathbb{K}} c_k^o T_k - \sum_{k \in \mathbb{K}} \sum_{w \in W_k^U} c_{kw}^U u_{kw}^U - \sum_{k \in \mathbb{K}} \sum_{i \in \mathbb{N}} \sum_{j \in \mathbb{N}, j \neq i} \gamma_{kij} d_{kij} \quad (1)$$

$$\sum_{m \in M} \alpha_{kmpi} \leq \sum_{w \in W_k^U} \sum_{j \in J_p} \sum_{s \in S} d_{pij} x_{pjkws}^U, \forall k \in \mathbb{K}, \forall p \in \mathbb{P}, \forall i \in \mathbb{I}_p \quad (2)$$

$$\sum_{w \in W_k^U} \left(w(x_{pj_1kw1}^U - x_{pj_2kw1}^U) + (2W_k^U - w)(x_{pj_1kw2}^U - x_{pj_2kw2}^U) \right)$$
$$+ 2W_k^U \left(\sum_{w \in W_k^U} \sum_{s \in S} x_{pj_2kws}^U - 1 \right) \leq 0, \forall k \in \mathbb{K}, \forall p \in \mathbb{P}^p, \forall j_1, j_2 \in \mathbb{J}_p, s_{pj_1j_2} = 1$$
$$(3)$$

$$\sum_{w \in W_k^U} \sum_{s \in S} x_{pj_2kw}^U \leq \sum_{j_1 \in \mathbb{J}_p} \sum_{w \in W_k^U} \sum_{s \in S} (x_{pj_1kw}^U s_{pj_1j_2}^U),$$
$$\forall k \in \mathbb{K}, \forall p \in \mathbb{P}, \forall j_2 \in \mathbb{J}_p, i = 1, d_{pij_2} = 0 \quad (4)$$

$$\sum_{w \in W_k^U} \sum_{s \in S} (x_{pj_1kws}^U + x_{pj_2kws}^U) \leq 1,$$
$$\forall k \in \mathbb{K}, \forall p \in \mathbb{P}, \forall j_1, j_2 \in \mathbb{J}_p, r_{pj_1j_2} = 1 \quad (5)$$

$$\beta_{kn} \geq \alpha_{kmpi}, \forall k \in \mathbb{K}, \forall m \in M, \forall p \in \mathbb{P}, \forall i \in \mathbb{I}_p, n \in \mathbb{N} \quad (6)$$

$$\sum_{j \in \mathbb{N}, i \neq j} \gamma_{kij} = \beta_{ki}, \forall k \in \mathbb{K}, \forall i \in \mathbb{N} \quad (7)$$

$$\sum_{i \in \mathbb{N}, i \neq j} \gamma_{kij} = \beta_{kj}, \forall k \in \mathbb{K}, \forall j \in \mathbb{N} \quad (8)$$

$$u_{ki} - u_{kj} + (N-1)\gamma_{kij} \leq N - 2, \forall k \in \mathbb{K}, \forall i, j \in \mathbb{N}, i \neq j \quad (9)$$

The objective function (1) maximizes the total profit in the recycling and remanufacturing process. Specifically, the first term captures revenue from selling disassembled subassemblies or components; the second term accounts for disassembly costs on U-shaped lines; the third and fourth terms represent the operating costs of turning on factories and activating workstations, respectively; and the final term reflects drone-based distribution costs. Constraint (2) uses decision variable α_{kmpi} to count the disassembled subassemblies. Constraint (3) enforces the precedence relationship between tasks, while (4) addresses tasks without predecessors at the initial stage. Constraint (5) ensures task assignments respect conflict constraints. Constraint (6) defines β_{kn} to count the manufacturing factories each drone must visit. Constraints (7) and (8) require each destination to have one incoming and one outgoing route. Constraint (9) adopts the MTZ formulation to eliminate subtours.

3 Proposed Algorithm

This work constructs a simulation environment for the remanufacturing optimization problem involving multiple products and factories, and models the environment based on the OpenAI Gym framework. The environment integrates multiple factors, including product disassembly, resource allocation of factories and workstations, and transportation path planning, and introduces drone-based distribution to reflect cost and benefit relationships in actual production and logistics processes. The core objective of the environment design is to provide reasonable and stable state, action, and reward signals for the improved Dueling DQN algorithm, thereby guiding the agent to achieve global profit maximization in complex decision problems.

To optimize the learning and convergence of the algorithm, the environment is designed in three main stages. An action mask is introduced to constrain the action space, reduce the probability of selecting invalid actions, improve learning efficiency, and simplify the reward function. In the first stage, the remanufacturing center assigns products to the disassembly lines of each disassembly factory. During this process, constraints (4) and (5) are considered. The status of each disassembly line being turned on is recorded, and the product selection space is adjusted dynamically. Proper product allocation optimizes resource utilization and supports subsequent disassembly tasks. In the second stage, the disassembly process is abstracted as an action sequence, where one disassembly task and its corresponding workstation position are selected at each step. Executable disassembly tasks are filtered based on the disassembly relationship matrix d_{pij}, and the present or subsequent workstations are selected based on the present position. In the third stage, after each disassembly task is completed, the maximum profit is calculated based on the obtained subassemblies and transportation cost. Since the number of nodes for drone distribution is relatively small, a dynamic programming algorithm is used to efficiently obtain the optimal distribution path and transportation cost. Based on the calculated profit, a reward function is designed and provided to the algorithm to optimize the learning process.

3.1 State Space

This design provides sufficient contextual information for the Dueling DQN algorithm, supporting more accurate value estimation and decision-making. The meanings of each dimension are as follows:

State Index 0. Product index P, denoting the currently processed product.

State Index 1. Factory index K, indicating the selected disassembly factory.

State Index 2. Workstation index W, representing the workstation assigned for the disassembly task.

State Index 3. Task index J, specifying the selected disassembly task.

State Indexes 4–8. Economic indicators accumulated in the system, including component revenue (profit from disassembled parts), disassembly cost (based on task duration and unit price), factory cycle cost (operating cost linked to the load on activated workstations), workstation activation cost (fixed and variable costs for enabling workstations), and transportation cost.

This state representation incorporates resource allocation details pertinent to the existing decision and cost and revenue metrics essential for reward calculation, thus providing rich contextual support that enhances the Dueling DQN's learning efficiency and convergence speed.

3.2 Action Space

The environment employs a discrete action space with an initial size of 50 actions. In the product allocation phase, actions are dedicated to factory selection through index-to-factory mapping via modulo operation. During task execution phases, each action represents a workstation-task combination (W, J). The environment dynamically generates executable actions through the algorithm to ensure validity, incorporating special commands (-1) for product switching or episode termination.

This hierarchical design strategically decouples product allocation from task execution decisions, reducing both action space dimensionality and decision complexity. The action mask in the algorithm dynamically generates valid actions by analyzing the present product P, active factory K, disassembly line L, and last workstation W. It systematically examines subassembly lists and the disassembly relationship matrix d_{pij} (where $d_{pij} = -1$ indicates executable tasks) to identify feasible tasks. Workstations on both active sides become available for U-shaped disassembly lines when one side is activated. All valid workstation-task pairs are combinatorially encoded, with the mandatory inclusion of the (-1) termination command when no tasks remain.

3.3 Reward Function

The reward function is designed to provide clear profit-based learning signals during disassembly operations while avoiding misleading incentives during non-productive stages. Specifically:

- **Zero-reward stages**: During product allocation, product switching, and terminal states, actions don't directly generate economic value. Thus:

$$R_t = 0 \quad \text{(non-disassembly operations)}$$

- **Profit-driven disassembly**: During disassembly operations, rewards reflect incremental economic gains:

$$R_t = \Delta P_t = P_t - P_{t-1} \tag{10}$$

where P_t denotes profit at timestep t and ΔP_t represents immediate profit change.

The designed reward formulation ensures three critical properties for effective reinforcement learning: (1) **Temporal alignment** between rewards and value-creating actions, guaranteeing that disassembly decisions receive immediate feedback proportional to their economic impact; (2) **Precise credit assignment** that directly attributes profit changes to specific operational choices, eliminating ambiguity in action valuation; (3) **Significant noise reduction** through the elimination of spurious reward signals during non-productive setup and transition phases, which could otherwise obscure the learning signal. This triad of features collectively enables stable policy convergence while maintaining economic interpretability.

3.4 Pseudocode

The Dueling DQN algorithm (Algorithm 1) implements an iterative agent-environment interaction loop. Initialization instantiates the dueling Q-network Q_θ (with separate V and A streams) and target network $Q_{\theta'}$. Each episode begins by resetting to initial state s_0 corresponding to the product allocation phase. At each timestep: 1. Action selection employs ϵ-greedy exploration constrained by action masking 2. Environment execution yields profit-driven rewards r_t and next state s_{t+1} 3. Transitions are stored in replay buffer $\mathcal{B}$

When $\mathcal{B}$ reaches capacity: 1. Minibatches are sampled for TD learning 2. Target values y_j are computed via $Q_{\theta'}$ 3. Online estimates q derive from Q_θ 4. Network parameters θ are updated via MSE minimization 5. Target network parameters θ' undergo soft synchronization 6. Exploration rate ϵ decays progressively

This cyclic process continues until terminal states, with the dual-stream architecture learning value estimates that maximize cumulative disassembly profit under environmental constraints.

Algorithm 1. Dueling DQN Algorithm

1: **Input:** State dim d_s, action dim d_a, learning rate α, discount γ, soft update τ
2: **Initialize:**
3:    Q-network Q_θ with dueling architecture (V and A streams)
4:    Target network $Q_{\theta'}$ ($\theta' \leftarrow \theta$)
5:    Replay buffer $\mathcal{B}$ with capacity N
6:    Exploration rate $\epsilon = 1.0$
7: **for** episode $= 1$ to M **do**
8:    Reset environment, observe state s_0
9:    **for** $t = 0$ to T **do**
10:      Choose action $a_t = \begin{cases} \text{random action} & \text{with prob } \epsilon \\ \arg\max_a Q(s_t) & \text{otherwise} \end{cases}$
11:      Execute a_t, observe r_t, s_{t+1}, d_t
12:      Store transition $(s_t, a_t, r_t, s_{t+1}, d_t)$ in $\mathcal{B}$
13:      **if** $\mathcal{B}$ has enough samples **then**
14:         Sample mini-batch $(s_j, a_j, r_j, s_{j+1}, d_j) \sim \mathcal{B}$
15:         Compute target:
16:         $V', A' \leftarrow Q_{\theta'}(s_{j+1})$
17:         $Q' \leftarrow V' + A' - \text{mean}(A')$
18:         $y_j = r_j + \gamma(1 - d_j)\max Q'$
19:         Compute current Q-value:
20:         $V, A \leftarrow Q_\theta(s_j)$
21:         $Q_j \leftarrow V + A - \text{mean}(A)$
22:         $Q(s_j, a_j) \leftarrow Q_j[a_j]$
23:         Update θ to minimize $\|y_j - Q(s_j, a_j)\|^2$
24:         Soft update: $\theta' \leftarrow \tau\theta + (1-\tau)\theta'$
25:         Decay ϵ
26:      **end if**
27:    **end for**
28: **end for**

4 Experimental Result and Analysis

To validate the model's correctness and the algorithm's effectiveness, the experimental case is first solved using the IBM ILOG CPLEX general solver to obtain the optimal solution. The same case is then solved using the Dueling DQN algorithm, and the results are compared for verification. Experiments are conducted on a Windows 11 system with an AMD Ryzen 7 4800H CPU (2.90 GHz, 16.00 GB RAM) and a GTX 1650 GPU.

This work considers four types of products for disassembly: washing machine, computer, clutch, and radio. The washing machine and computer represent small-scale disassembly tasks, while the clutch and radio involve more complex disassembly processes. Specifically, the clutch features intricate precedence relationships, and the radio comprises more subassemblies, increasing disassembly difficulty. To simplify modeling assumptions, the number of disassembly factories is fixed at 2, and the number of manufacturing factories is fixed at 3. Disassembly costs, subassembly selling prices, and transportation costs are randomly generated based on normal distributions centered on predefined baseline values, simulating variability under different market conditions. Each disassembly case involves a unique product combination. Table 1 presents detailed product information and case configurations.

Table 1. Products for disassembly test

Case ID	Number of product				Number of disassembly tasks
	Washing machine	Computer	Clutch	Radio	
1	1	0	0	0	13
2	1	1	0	0	26
3	1	1	1	0	33
4	1	1	1	1	63
5	2	1	1	1	76
6	2	2	1	1	89

To evaluate the effectiveness of the Dueling DQN algorithm, benchmark cases are solved using CPLEX with a maximum runtime of 3 h. DQN algorithm from the Stable Baselines framework is also used as comparative methods and trained with default parameters. Figure 2 illustrates the profit trends during training across different cases.

Experimental results show that Dueling DQN consistently achieves the highest or near-highest convergence values in most cases and continues to improve profits in later training stages, demonstrating strong convergence capability. As the base version, DQN exhibits significant oscillations during training; while it occasionally achieves high profits, its overall stability and performance are inferior to Dueling DQN. Overall, the Dueling DQN algorithm exhibits superior profit optimization and stable convergence in most scenarios, demonstrating its suitability for the multi-factory remanufacturing optimization problem addressed in this work.

Dueling DQN decomposes the Q-network into two streams: one estimates the state-value function, and the other estimates the advantage function $A(s,a)$, representing the relative importance of each action in a given state. This architecture enables a more accurate estimation of state values and action advantages, leading to faster and more stable convergence toward the optimal policy. The charts show that Dueling DQN often achieves higher objective values in later training stages, suggesting improved action evaluation and stronger policy refinement compared to other algorithms. Dueling DQN consistently delivers strong or optimal performance across most test environments, highlighting its advantage in action-value estimation and effectiveness in deep reinforcement learning tasks.

Table 2 reports the objective values obtained by CPLEX and the maximum values achieved by each algorithm. The GAP column reflects the difference

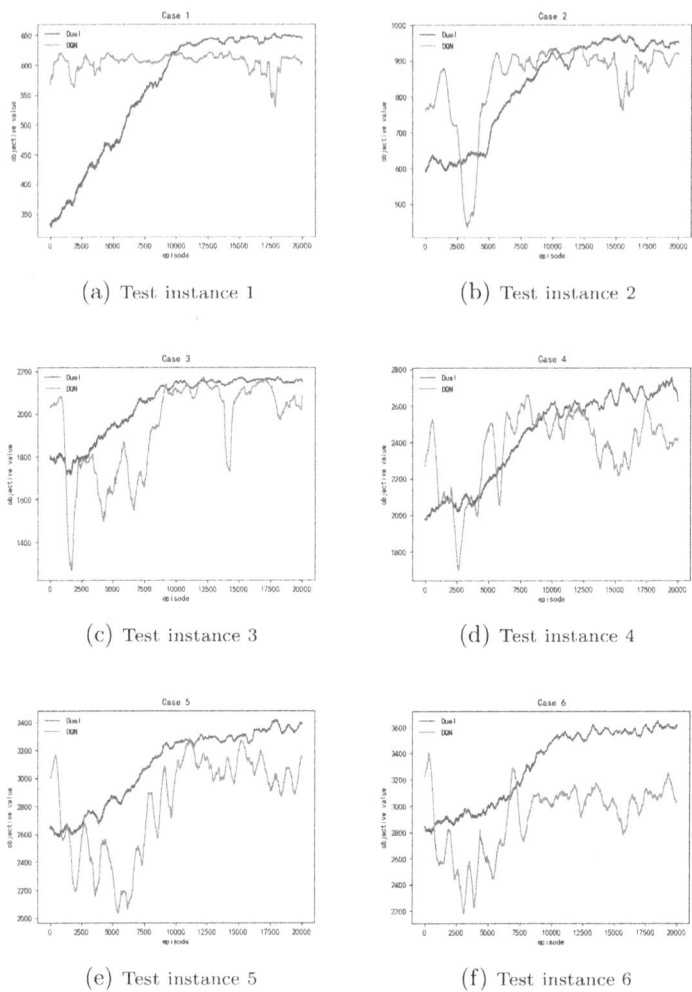

Fig. 2. A comprehensive comparison of the speed of profit convergence with the final gain of Dueling DQN compared to standard DQN in six test instances

between algorithm results and the optimal solutions provided by CPLEX. In Case 6, the CPLEX solver fails to return a solution within the 3-hour time limit. In Cases 1 to 5, Dueling DQN yields results closest to the CPLEX benchmarks and outperforms DQN. Moreover, in complex cases, it exhibits lower performance fluctuations. By decoupling state-value and advantage estimations, Dueling DQN enhances action-value precision, achieving results closer to CPLEX's global optimum—particularly in environments with large state spaces and high decision complexity.

Table 3 further lists the comparison of running times for each algorithm. In terms of computational efficiency, Dueling DQN and DQN, as value-based

Table 2. CPLEX and the objective function values obtained by each algorithm for solving MUD-MROP

Case ID	Objective function value			
	Dueling-DQN	DQN	CPLEX	GAP
1	664	664	664	0.00%
2	1077	1055	1082	0.46%
3	2366	2357	2417	2.11%
4	3105	2988	3221	3.60%
5	3884	3772	4086	4.94%
6	4075	4074	-	-

Table 3. The training time of each algorithm

Case ID	Solution time		
	Dueling-DQN	DQN	CPLEX
1	114.13	305.46	4.73
2	198.65	416.77	5.35
3	298.74	545.15	6.65
4	443.39	646.29	6.03
5	512.94	747.64	2.80
6	582.43	868.05	10800.00

methods, require fewer parameter updates during neural network training, thus demonstrating faster computation speeds. Although Dueling DQN's computation time is higher than CPLEX when solving small-scale problems, it is significantly lower than DQN, and its runtime remains manageable for medium and large-scale problems.

5 Conclusion

This study presents the multi-factory remanufacturing optimization incorporating drone delivery, establishing a reinforcement learning environment and solving the problem using the Dueling DQN algorithm. Experimental results demonstrate that the Dueling DQN achieves superior solution quality and generalization capability in most test cases, significantly outperforming the standard DQN. Notably, under high problem complexity, it maintains robust performance, confirming its advantages in state-value estimation and action selection. However, the model has certain limitations. The drone delivery stage assumes a single drone per factory operating in static environments, without considering real-world factors such as dynamic weather conditions, airspace restrictions, battery degradation, or multi-drone coordination. Future research will expand this framework to multi-objective optimization, aiming to balance profit maximization with environmental sustainability in multi-factory remanufacturing [9,23–26]. We will also explore explainable AI approaches for problem-solving [27].

Acknowledgments. This research was funded by 'National Local Joint Engineering Laboratory for Optimization of Petrochemical Process Operation and Energy saving Technology' grant number LJ232410148002. and 'the Innovation Team Project of the Educational Department of Liaoning Province' grant number LJ222410148036.

References

1. Guo, X., Wei, T., Wang, J., Liu, S., Qin, S., Qi, L.: Multiobjective u-shaped disassembly line balancing problem considering human fatigue index and an efficient solution. IEEE Trans. Comput. Social Syst. **10**(4), 2061–2073 (2022)
2. Guo, X., et al.: Human–robot collaborative disassembly line balancing problem with stochastic operation time and a solution via multi-objective shuffled frog leaping algorithm. IEEE Trans. Autom. Sci. Eng. **21**(3), 4448–4459 (2024). https://doi.org/10.1109/TASE.2023.3296733
3. Zhao, Z., Jiang, Q., Liu, S., Zhou, M.C., Yang, X., Guo, X.: Energy, cost and job-tardiness-minimized scheduling of energy-intensive and high-cost industrial production systems. Eng. Appl. Artif. Intell. **133**, 108477 (2024). https://doi.org/10.1016/j.engappai.2024.108477
4. Guo, X., Zhang, Z., Qi, L., Liu, S., Tang, Y., Zhao, Z.: Stochastic hybrid discrete grey wolf optimizer for multi-objective disassembly sequencing and line balancing planning in disassembling multiple products. IEEE Trans. Autom. Sci. Eng. **19**(3), 1744–1756 (2021)
5. Guo, X., Zhou, M., Liu, S., Qi, L.: Lexicographic multiobjective scatter search for the optimization of sequence-dependent selective disassembly subject to multiresource constraints. IEEE Trans. Cybernet. **50**(7), 3307–3317 (2019)
6. Mete, S., Çil, Z.A., Özceylan, E., Ağpak, K., Battaïa, O.: An optimisation support for the design of hybrid production lines including assembly and disassembly tasks. Int. J. Product. Res. **56**(24), 7375–7389 (2018). https://doi.org/10.1080/00207543.2018.1428774
7. Hanafy, M., ElMaraghy, H.: Modular product platform configuration and co-planning of assembly lines using assembly and disassembly. J. Manuf. Syst. **42**, 289–305 (2017)
8. Pal, A., Restrepo, V., Goswami, D., Martinez, R.V.: Exploiting mechanical instabilities in soft robotics: control, sensing, and actuation. Adv. Mater. **33**(19), 2006939 (2021)
9. Qin, S., et al.: An optimized advantage actor-critic algorithm for disassembly line balancing problem considering disassembly tool degradation. Mathematics **12**(6), 836 (2024)
10. Qin, S., Li, J., Wang, J., Guo, X., Liu, S., Qi, L.: A salp swarm algorithm for parallel disassembly line balancing considering workers with government benefits. IEEE Trans. Comput. Social Syst. **11**(1), 282–291 (2023)
11. Zeng, Y., Zhang, Z., Yin, T., Zheng, H.: Robotic disassembly line balancing and sequencing problem considering energy-saving and high-profit for waste household appliances. J. Clean. Prod. **381**, 135209 (2022)

12. Li, C., Guo, X.W., Wang, J., Qin, S., Qi, L., Tang, Y.: Grey wolf algorithm for human-robot collaborative disassembly line balancing problem subject to dangerous components. In: 2022 IEEE International Conference on Networking, Sensing and Control (ICNSC). IEEE, 2022, pp. 1–6 (2022)
13. Kozuno, T., et al.: Revisiting Peng's Q (λ) for Modern Reinforcement Learning. In: International Conference on Machine Learning. PMLR (2021)
14. Bahubalendruni, M.R., Varupala, V.P.: Disassembly sequence planning for safe disposal of end-of-life waste electric and electronic equipment. National Acad. Sci. Letters **44**(3), 243–247 (2021)
15. Guo, X., Zhou, M., Abusorrah, A., Alsokhiry, F., Sedraoui, K.: Disassembly sequence planning: a survey. IEEE/CAA J. Automatica Sinica **8**(7), 1308–1324 (2020)
16. Gungor, A., Gupta, S.M.: A solution approach to the disassembly line balancing problem in the presence of task failures. Int. J. Prod. Res. **39**(7), 1427–1467 (2001)
17. Fu, Y., Zhou, M., Guo, X., Qi, L.: Stochastic multi-objective integrated disassembly-reprocessing-reassembly scheduling via fruit fly optimization algorithm. J. Clean. Prod. **278**, 123364 (2021)
18. Guo, X., et al.: Modeling and optimization of multiproduct human–robot collaborative hybrid disassembly line balancing with resource sharing. IEEE Trans. Comput. Social Syst., 1–16 (2025). https://doi.org/10.1109/TCSS.2025.3540070
19. Guo, X., Chen, L., Qi, L., Wang, J., Qin, S., Chatterjee, M., Kang, Q.: Multifactory disassembly process optimization considering worker posture. IEEE Trans. Comput. Social Syst. 1–13 (2025). https://doi.org/10.1109/TCSS.2025.3540565
20. Wei, T., et al.: A multiobjective discrete harmony search optimizer for disassembly line balancing problems considering human factors. IEEE Trans. Human-Mach. Syst. **55**(2), 124–133 (2025). https://doi.org/10.1109/THMS.2025.3528629
21. Gu, J., Guo, Z., Wang, J., Qi, L., Qin, S., Zhang, S.: Optimization of multifactory remanufacturing processes with shared transportation resources using the alns algorithm. Int. J. Artif. Intell. Green Manufact. **1**(1) (2025)
22. Qin, S., et al.: Expanded discrete migratory bird optimizer for circular disassembly line balancing with tool deterioration and replacement. Int. J. Artif. Intell. Green Manufact. **1**(1) (2025)
23. Ji, Y., Zhao, Z., Liu, S., Yong, X.: Machine learning-based prediction of surplus material in intelligent production processes. Int. J. Artif. Intell. Green Manufact. **1**(1) (2025)
24. Yan, S., Guo, H., Liu, S.: A partition stacking classification framework with oversampling for quality prediction of aluminum alloy ingots. Int. J. Artif. Intell. Green Manufact. **1**(1) (2025)
25. Qi, L., Liang, L., Luan, W., Lu, T., Guo, X., Talukder, Q.T.A.: Integrated control strategies for freeway bottlenecks with vehicle platooning. Int. J. Artif. Intell. Green Manufact. **1**(1) (2025)
26. Rahkar Farshi, T.: Battle royale optimization algorithm. Neural Comput. Appl. **33**(4), 1139–1157 (2021)
27. Han, H., Li, W., Wang, J., Qin, G., Qin, X.: Enhance explainability of manifold learning. Neurocomputing **500**, 877–895 (2022)

Author Index

A
Amponsah, Sheilla 125

C
Chen, Yaodong 107
Choupani, Roya 143

D
Dai, Shaokang 184
Ding, Tianqi 3
Dogdu, Erdogan 143
Dong, Liang 3

G
Gao, Yuanfan 92
Guo, Enyan 92
Guo, Xiwang 184

H
Han, Ashley 16
Han, Henry 16, 39
Huang, Heming 77
Hubbard, Keith 125

J
Ji, Yingjun 184

K
Kahle, David 58
Kolli, Sumanth 39

L
Le, Minh 143

M
Miyakawa, Evan 58

Q
Qin, Shujin 184

S
Schubert, Keith E. 3
Shaw, Lily 143
Shi, Runqing 107
Song, Yan 107

W
Wang, Jiacun 184
Wang, Shuo 107
Wei, Yuanyuan 77
Womack, Steven 143
Wu, Bella 184
Wu, Chujiang 39

X
Xiang, Dawei 3
Xiao, Jianwei 163
Xue, Yanyan 163

Y
Yang, Xuemei 163

Z
Zhang, Fen 92
Zhang, Yanbang 92
Zhang, Yuting 107
Zheng, Yuanyuan 163
Zhou, Mengshan 107
Zhou, Yao 107

Made in the USA
Monee, IL
03 May 2026

49438661R00118